BIOFI

Methods for Enzymatic Release of Microorganisms

Jean F. Brisou
National Corresponding Member of
the Academy of Medicine

CRC Press
Taylor & Francis Group
Boca Raton London New York

CRC Press is an imprint of the
Taylor & Francis Group, an **informa** business

First published 1995 by CRC Press
Taylor & Francis Group
6000 Broken Sound Parkway NW, Suite
300 Boca Raton, FL 33487-2742

Reissued 2018 by CRC Press

A Library of Congress record exists under LC control number: 95014237

Publisher's Note
The publisher has gone to great lengths to ensure the quality of this reprint but points out that some imperfections in the original copies may be apparent.

Disclaimer
The publisher has made every effort to trace copyright holders and welcomes correspondence from those they have been unable to contact.

ISBN 13: 978-1-138-50665-7 (hbk)
ISBN 13: 978-1-138-55770-3 (pbk)
ISBN 13: 978-1-315-14999-8 (ebk)

Visit the Taylor & Francis Web site at http://www.taylorandfrancis.com and the CRC Press Web site at http://www.crcpress.com

PREFACE

This monograph presents a new technique for detecting microorganisms, more specifically bacteria, that populate all the levels of the biosphere. Since the majority of these microorganisms, constructors of biofilms and agglomerates, are attached to living or inert surfaces, they are not detected by the usual investigative methods. After gaining more knowledge of the mechanisms and stages of adherence of these microbes on the host sites, which include interfaces of all kinds, we attempted to "release" them from their habitats using enzymatic means that are as nontraumatic as possible. The reasons behind this process are presented in this book, which is broken down into three parts:

Part I, which is devoted to the mechanisms of adherence, includes two chapters. Chapter 1 is devoted to membrane structures, adherence stages, and attachment organs, known by the name of adhesins. A paragraph summarizes a few concepts on genetics concerning the biosynthesis of these anchorage elements. Chapter 2 exclusively concerns the biochemical aspects of this adherence. Presented in the form of helpful reminders, this chapter reviews a few notions of biochemistry, and it is intentionally limited to the essential knowledge needed for the comprehension and justification of techniques used for release and for the understanding of the life of germs in nature.

Part II directly picks up with microbial ecology with a glance at the bacterial population of tissues as well as land and aquatic microbiocenoses in general. Numerous documents illustrate the relations maintained between microorganisms and nature, the importance of colonizations, the construction of biofilms, and the formation of aggregates.

Part III is specifically devoted to technique. What we know now about the mechanisms of adherence justifies the choice of the techniques suggested for releases using wisely chosen enzymes which sometimes act on the host surfaces and sometimes on the adherence organs. The experimental findings recorded since 1979 by us and by a few others are summarized most often in tables followed by a few remarks.

The postface comments on these results and draws some prospects for applications that suit a maximum number of situations, such as applications in nature, medicine, environmental hygiene, the foodstuffs industry, food monitoring, etc. These methods are still poorly known, but they deserve, due to the results obtained, wider acknowledgment and further testing. The possible applications are infinite.

ACKNOWLEDGMENTS

All the experiments related here were performed in the laboratory of the Military Teaching Hospital in Toulon. I am especially grateful for the welcome given by the authorities and the effective cooperation of the department heads in bacteriology, Bernard Brisou and J.Cl. Roche, and that of the head pharmacist, M. Lecarpentier, in charge of the electron microscopy department. I would like to take this opportunity to express my thanks.

I would also like to mention Professor Edouard Boureau, who was so kind as to present my initial results to the Academy of Sciences (Paris) and to whom I am deeply grateful. I can naturally not mention all my colleagues, friends, and assistants who participated in my "Bacteria Hunts" in a wide range of conditions over the 65 years of my scientific career. However, as far as canvassing extreme environments is concerned, I readily thank P. Niaussat, with whom I participated in the bacterial enumeration of the extraordinary Clipperton atoll in the Pacific Ocean, and M. Courtois with whom I worked at the most extreme hypersalted lake in the world, Lake Assal, located in the Affar-Issa territory.

I would also like to express my thanks to the colleagues who directly participated in the publication of this monograph with their support and encouragement: the professors at the University Hospital Centers, P. Ambroise Thomas (Grenoble), F. Denis (Limoges), J.L. Jacquemin (Poitiers), and J. Paccalin (Bordeaux); the Doctor General L. Force of the National Navy; and Professor Richelle-Maurer and Z. Moureau of the Royal Institute of Brussels. Also, for the very longstanding and constant collaboration of Professor R. Moreau and the veterinarian C. Tysset, I express my warmest thanks. To Professor R. Vargues of the University Hospital Center of Tours, a constantly understanding friend who readily accepted my undertakings, I owe a great deal and would like to take this opportunity to express my admiration. And finally, to the memory of Professor R. Babin to whom I am bound by approximately 40 years of joint work.

On the technical level, as far as the production of this monograph is concerned, I would like to express my warm thanks to Leslie Marcus who took care of the translation, and to my nephew Jean-Pierre Gallacier for his precious help with the documentation of this work, and to my son Jean Philippe Brisou and James Cossais who took care of the illustrations.

CONTENTS

PART II
BACTERIA LIFE IN THE WILD

CHAPTER 1
ADHERENCE IN LIVING TISSUES

1. Cellular Receptors ... 83

CHAPTER 2
ADHERENCE TO A FEW ORGANS AND ORGANISMS

1. Man and Animals ... 91
    1.1. Microbisms of the Mouth and Teeth ... 91
    1.2. Microbisms of the Digestive Tube ... 95
    1.3. Microbisms of the Stomach ... 96
    1.4. Microbisms of the Urogenital Apparatus ... 100

CHAPTER 3
NATURAL MICROBIOCENOSES

1. Vegetals ... 105
2. The Rhizosphere ... 106
3. Soils and Sediments ... 110
4. Aquatic Environments ... 113
5. Gas/Liquid Interfaces ... 120
6. Nonmiscible Liquid Interfaces ... 120
7. Microbial Activity on Interfaces ... 122
8. Role of Plankton in Aquatic Environments ... 124
9. Polymorphism of Bacteria in Nature, Agglomerates ... 125
10. Inert Surfaces ... 132
11. Conclusion ... 137

PART III
ENZYMATIC RELEASE TECHNIQUES OF BACTERIA

CHAPTER 1
BASIC TECHNIQUES

1. Introduction ... 141
2. Attack of Host Sites or Substrates ... 142
    2.1. Inorganic or Inert Surfaces ... 142
    2.2. Organic Surfaces ... 142
    2.3. Analysis of North Atlantic Seawater ... 143

DEDICATION

To my grandson, Patrick Brisou

INTRODUCTION

INTRODUCTION

Bacteria, the first acknowledged signs of cellular life, left the print of their walls in the sediments they lived in approximately 3,600,000,000 years ago. "Life was already beginning to play a role of utmost importance in the changes of the face of the earth," wrote the French paleontologist Marcellin Boule (1863–1942). Microorganisms — bacteria, viruses, yeasts, lower fungi, protists — do indeed attach themselves very rapidly onto inert or living media, forming aggregates of an extreme complexity. As a result, a large part of the attached microbial populations elude conventional investigation techniques. To offset this capital cause of error in reading the book *Nature* in 1979, I suggested releasing these microbes from their shelters, "flushing them out", through a nontraumatic enzymatic pathway, in order to make their culture and determination easier, either by acting directly on their attachment elements or by destroying their media.

Bacteria are the only microorganisms discussed here, based on my personal experience, but it is obvious that all the general concepts presented in this book are applicable to viruses and all the other microorganisms. The reality of this adherence dominates microbial ecology, calling for a complete review of out-dated concepts. Other than a few rare exceptions, there are practically no free bacteria in nature. Many years of experience and numerous observations have made it possible to suggest a new strategy for studying environmental microbiocenoses as well as animal and plant microbisms.

As early as the 18th century, it was quickly understood that the attachment, or immobilization, of unicellular beings was related to the surfaces, volumes, and porosity of materials. "Microbists" came to the still-valid conclusion that they were the same capillary forces that retain the organic substances and germs, i.e., the food matter and the microscopic beings that feed off it in the top layers of soils (Duclaux). This knowledge has been checked again and again, and in his microbiology treaty of 1898, Duclaux[1] insists on the close bonds uniting organic matter and microbial life. Although the long survival of microbes in the environment was already acknowledged then, the physiochemical, molecular, and genetic mechanisms of this survival remained, for the most part, unknown.

This survival depends on three requirements, applicable to all living beings:

- Protection (safety)
- Food (energy supply)
- Reproduction possibilities (growth)

In the past, adsorption was acknowledged as being practically the only important factor enabling the survival of microorganisms in nature. The substrates or "interfaces" on which "biofilms" are formed were studied extensively. Although the two components of the attached/attacher (or sensor) relation cannot be dissociated, explanations based purely on physics are no longer sufficient.

The concept of life and survival will not be considered here to answer the question "why", which would lead us to the thoughts of philosphers and finalists, but the question "how", which will keep us in the field of observation, experimentation, and description.

Microbiologists do not study beings in the linear sense of the term, but in populations of billions of cells that, in the optimal conditions of a given environment are divided, on the average, every 30 minutes! In conditions such as these, the researcher only records "snapshots" and finds himself or herself, like a specialist of quantum physics, in the field of probabilities. At the heart of very complex bacterial populations, constant, beneficial, or harmful exchanges take place, which were effectively analyzed by Alexander[2] in 1971.

Genetics pursues its task enabling transformations by DNA, conjugations, phagic conversions, and transductions; plasmids and episomes go from one cell to another, but at rhythms such that it would be careless to assimilate to what we observe *in vitro* with "tamed" bacterial strains. It may be assumed that the "free" microbial populations are located at a "high-risk" level. They are in danger unless they have particular forms of resistance, such as spores resistant to aggressions. Others can adapt to prolonged fast and low maintenance rations. Such very unequally distributed aptitudes play a role in ecological valency. Although certain bacteria are extremely fragile, others, on the contrary, resist extreme environments, the study of which is especially interesting (Gould and Corry, 1980).[3]

Bacteria can enter into lethargy, or sleep, and play "Sleeping Beauty". We devoted a certain number of studies (1960–1969)[4] to these filterable, dwarf forms that we consider as "hidden forms", since they elude the usual detection techniques. Thanks to certain devices, they recover their normal form, their antigenicity, and their experimental pathogenic power. Some authors would rather consider them as "debilitated" bacteria.

Over the past 25 years, the biochemical mechanisms of adherence have been s1pecified. Some researchers, performing a genuine return to the past, evidenced considerable differences between the behavior of bacteria *in vitro* and in nature. In 1980, Roth[5] resumed studies concerning carbohydrates serving as a protection to bacteria living in water. Sutherland[6] confirmed the universality of the anchorage fibers that were named "Glycocalyx". Their polysaccharide nature was acknowledged. Interest was then focused on other attachment organelles which will be examined in detail under the name of "adhesins". Costerton and his colleagues[7,8] widely contributed to the spread of this fundamental knowledge. Other authors will be mentioned throughout this monograph, the goal of which is to bring together the data related to the mechanisms of bacteria attachment on a wide range of media so as to gain practical knowledge from it. The reader will learn technical means enabling the implementation of a new hunting strategy for "wild bacteria", irrespective of the environment or organism considered.

"Enzymatic Release" is the very basic foundation of this strategy. It questions a great deal of data considered as reference "standards". Released bacteria do cultivate more quickly. They become more sensitive to means of defense, and their sensitivity to antibacterial substances is increased, sometimes by 50%. This technique enables the isolation of species undetectable by conventional methods. Bacterial numerations will always be inaccurate so qualitative determinations take priority, without excluding the others. Current technology implements methods based on the amplification, by polymerized chain reactions, of nucleotidic sequences, called "PCR". This technique is guaranteed deserved success since it enables the quick detection of a microorganism, bacteria, virus, or even parasite, which are all undesirable in food, a pathological product, or even in a specific environment. Epistemologically, it is another level of observation which cannot in any case exclude the isolation, accurate identification, or possible pathogenic power of the identified microorganism, and especially its behavior in the face of the usual antagonists.

Specialists acknowledge the advantages of these methods as well as their limits. However, they will not be debated here as their discussion would lead us too far away from our subject.

The bacterial cell, or prokaryote, is the smallest known living unit. In the protoplasmic mass lies a free, unique chromosome, constantly changing in shape, with which small genetic elements are associated and sometimes attached on this chromosome: they are episomes, sometimes free in the cytoplasm and known by the name of plasmids.

There is not a membrane separating the chromosome from the cytoplasm. This chromosome is itself attached to the cytoplasmic membrane by tubular or vesiculous invaginations — mesosomes — more developed in Gram-positive bacteria.

The cell is naturally rich in ribosomes, where proteins are synthesized. In *Escherichia coli* up to 18,000 are counted. These organella, composed of 2 subunits, sometimes form veritable strings named polyosomes. Rhapidosomes, a kind of small, extremely thin stick encountered in *Pseudomonas*, a certain number of enterobacteria and Photobacterium, etc., should also be mentioned. Their function is still rather poorly known. They are interpreted as being the remains of bacteriophages. They are attributed with lytic virtues while assimilating them to bacteriocins.

Lastly, vacuoles, sometimes rich in lipids, and various granulations should be mentioned. The focus here will only be on what directly concerns adherence, its consequences, and what leads to enzymatic release. This entire cytoplasmic mass, with its contents, is wrapped in a thin, fragile, double-layered membrane. It is all protected by a thick wall that will be studied in depth in Part I, Chapter 2. All the organs of adherence known by the name of adhesins are implanted on this wall.

PART I
Adherence in Microbiology

1   STRUCTURES AND MECHANISMS

1. DEFINITION

As adherence of all microorganisms onto inert or living interfaces has been acknowledged for a long time, the mechanisms of it will be summarized by taking into account current data, using bacteria as an example. The framework of this essentially technical monograph does not justify developments concerning extremely complex biochemistry which would be more suitable for a "Treaty".

The process of adherence takes place schematically in three stages. The scenario is generally completed in 20 to 30 min in the optimal conditions of the environment. This is naturally not an absolute rule, simply a rough idea. The process takes place as follows:

1. *Adsorption:* Adsorption falls within the realm of physical chemistry. Ionic forces play a substantial role.
2. *Adhesion:* Adhesion is specific and characterized by a stereochemical molecular recognition, but it is a reversible process.
3. *Adherence:* Irreversible, terminal stage.

This conventional outline is sometimes criticized. It nevertheless has the advantage of being simple and didactic. What essentially stands out is that the two actors to be taken into account in this scenario are:

- The "Attached" (bacteria, for example)
- The "Attacher" (the living or inert host substrate)

Before analyzing these two fundamental components, a few concepts concerning cellular membranes which envelope bacteria as well as host tissues should be reviewed. All living cells are limited by a membrane that plays an important role in adherence phenomena.

2. REVIEW OF MEMBRANE STRUCTURES

Numerous outlines published in various works (see bibliography) illustrate the complexity of membrane structures. Amphipolar molecules and phospholipids are superimposed in monomolecular layers. Certain internal asymmetries are responsible for variations in the distribution of electric charges and cause imbalances inducing the appearance of dipoles differentiated as instantaneous, permanent, or adipolar. All these elements obey the laws of Van der Waals, or are guided according to various modalities analyzed and debated by different authors. Invoking electrostatic forces, Keesen constructs singular associations and alignments; Debye, a partisan of electrostatic induction, is opposed to London, a partisan of electrokinetics and the action of apolar molecules.

This debate will be avoided by remembering that the Van der Waals type bonds, weaker than ionic covalents, are exceptions to the saturation phenomena and play an unquestionable role. This point will be dealt with later in the development of life processes. The interface phenomena that will be discussed later were perfectly analyzed in 1971 by Mazliak.[9] The spatial congestion characterizing the chains of unsaturated fat acids is more substantial than that occupied by the saturated fat acids. The space included between these alignments facilitates the $-CH_2$ bonds and ensures the rigidity of the whole. A monomolecular film can thus be formed. The polar heads contract bonds with various cations, notably the bivalents Ca^+ and Mg. They are united to remainders of PO_4H_3 themselves capable of belonging to two different but similar molecules. In this way, these bivalent cations ensure the bringing together of the polar heads and promote the organization of phospholipids in compact lamelar systems. Monovalent cations, such as Na^+ or K^+ only recognizing one molecule, cannot claim such unions.

2.1. Membrane Proteins

Membrane proteins are traditionally broken down into two groups: structural proteins and enzyme proteins intimately linked to the membranes.

2.1.1. Structural Proteins

Structural proteins are essentially hydrophobic — relatively unsoluble in a medium close to neutrality. Deprived of any enzymatic activity, they have the capacity of uniting with phospholipids and giving birth to complex compounds. In an alkaline medium, they associate to engender substances that sediment at low speed. In a neutral medium, the monomers aggregate and organize small-diameter fibrils. Mg^{++} facilitates the development of the process; for example, myeline and fongic mitochondrias can be given as proof.

2.1.2. Enzyme Proteins

Due to the very substantial number of enzyme proteins, they cannot be counted here. A few examples are sufficient throughout this study to confirm the importance of these enzyme proteins. Along the way, adenosine triphosphatase, 5'nucleotidase, phosphatase glycerol, and other phosphatases, ribonuclease, numerous oxidoreductases, phospholipase A, etc. will be encountered.

2.2. Renewal of Membrane Constituents

The cellular membrane is the theater of considerable biochemical activity. It ensures that order is maintained in the heart of the cell. Life can be interpreted as a permanent reordering, a "neguentropy". Membrane structural molecules accomplish their mission based on their specific, catalytic, enzymatic potentialities, subjected to constant renewals. Mazliak wrote that "Everything happens, in the cells, as if masons were constantly replacing the bricks of the cell walls with new bricks, to make sure the walls are always new." This interpretation, full of imagery, shows the differences recorded between the cells depending on their age, their physiological condition, and the conditions of the environment in which microorganisms live.

Everyone knows that in optimal growth conditions a bacterial cell, in the strict sense of the term, 'becomes' and is transformed from one second to the next, giving birth to two daughter cells. In this life teeming with bacterial populations, only "snapshots" are taken. From one instant to the next, the antigenic patterns — the formation or disappearance of flagella — appear or disappear in function with the conditions of the medium. Wide variations in enzymatic activity are also observed, modifying the behavior of bacteria in general. Basic constituents can undergo completely unpredictable, profound reorganizations. Certain authors express all these modifications by diagrams that control their imagination. There are some twenty-odd models, many of which are inspired by Danielli's[10] ideas, published between 1943 and 1956. Some invent micellar structures (Paysant et al.[11]). Kavanau[12] attempts to make things clearer by limiting everything to two stable models, one of which was qualified as "open" and the other considered as "closed". Like an evolved chemical body, the membrane goes, depending on the circumstances and the constraints of the medium, from one configuration to another.

Cellular membranes, in all living beings, ensure the control of the movement of components of the microenvironment on the one hand and that of the intracellular substances on the other hand. Far from being passive, they recognize and sort everything that can enter in the cytoplasm and exit out of it. Pure and simple permeability does not obey the laws of physics, and in all but a few exceptions, the cell is alive; it reacts in every sense of the word, "... without transgressing the limits of what is possible" (F. Jacob).

We are thus led to the intimate, equally complex mechanisms of permeation. Nevertheless, certain metabolized substances obey the laws of diffusion of dissolved products depending on the gradient of concentration, but the speed of this diffusion is often slow. The highly selective character of the membrane with respect to mineral ions is also sufficiently known. It should simply be recalled that although these movements can go against the gradient of concentration, it is difficult to establish general laws based on them. It is sometimes possible to observe active transportation during which the ionic flow runs in the opposite direction of the concentration gradient, thus confirming the above. Such a phenomenon requires an additional expenditure of the cell's energy.

The breakdown of biopolymers is subjected to their spatial configuration, to their structure as much as to the ionic composition of the immediate environment, therefore to the breakdown of electric charges, which are never uniform nor continuous. Glycoproteins are, for example, always charged and intervene significantly in intercellular reactions, notably on the receptors assimilated to genuine biological signals of high specificity. The major role of such receptors dominates the development of adherence in all microorganisms, including viruses.

To sum things up, cellular membranes are composed of an assembly of intrinsic proteins located inside a lipidic layer and of extrinsic proteins emerging at the surface. The molecular orientation of each of the intrinsic proteins is specific to the species. Oligosides and monosaccharides form chains or unite together directly with certain proteins to give birth to glycoproteins or lipids. Other lipids form two quantitatively and qualitatively different layers. This external position of carbohydrates opposing itself to that of extrinsic proteins attached on the internal side, therefore in direct contact with the cytoplasm, causes a membrane asymmetry which may explain certain behaviors during the attachment phenomena of foreign molecules.

The structure of cellular membranes varies as much in bacteria as in all living beings. Substantial variations should therefore be expected in the behavior of Gram-positive and Gram-negative bacteria, Mycobacteria, and others. Their behavior, as will be specified and demonstrated based on concrete examples, corresponds to these structural and biochemical variants. The results are sometimes surprising. The cells are therefore, as a whole, covered by a hairy wrapping composed of extremely fine cilia that play an important role in adherence. These structures are complex, stereochemical, and specific. The main constituents are rich in carbohydrates, glycopeptides, glycoproteins, and mucopolysaccharide acids. These viscous, gelantinous, mucous mucopolysaccharides provide the lubrication and cohesion of the cells.

A primordial, almost omnipresent element is none other than hyaluronic acid, which often prompted us to use hyaluronidase with frequent success in our work on enzymatic release. It is a polymer, the basic disaccharide of which is formed by D-glucuronic acid and N-acetyl-D-glucosamine linked in alpha (1–3). A close substance is chondroitin, also frequent on the cellular level, in which glucosamine is replaced by N-acetyl galactosamine. The bacterial walls

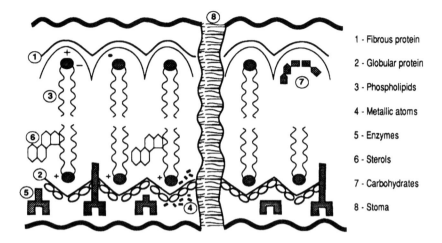

1 - Fibrous protein

2 - Globular protein

3 - Phospholipids

4 - Metallic atoms

5 - Enzymes

6 - Sterols

7 - Carbohydrates

8 - Stoma

Figure 1. Schematic representation of membrane.

contain mureins, heteropolyosides of muramic N-acetyl acid, and hexoseamine N-acetyl. In plant cells, a few polyosides dominate the whole. Once again, this is taken into account in choosing release enzymes. (See Figure 1.)

3. ADHERENCE MECHANISMS

Direct observation of microorganisms using optical microscopes, either alive or after various colorations, remains the most common, but perfunctory practice. It is still practically the same as that used by "microbists" of the 17th and 18th centuries. Electron microscopy, reserved for specialized studies, provides access to an entirely different level of observation. It reveals that germs, extremely varied in form, are in fact covered with "clothing" made up of fimbriae, glycocalyx, capsules, pili, flagella, and other accessories which make them hairy, villous beings as numerous documents published over the past 20 years show, some of which could pass for genuine works of art. Such is the reality that justifies the idea of an "Ecto-biology" suggested by Vosbeck and his colleagues[13] in 1981.

To clarify and simplify a relatively complex presentation, as well as to avoid "dilutions" and repetitions, an outline identical to one that would be followed for biochemistry has been adopted.

Adhesins are the elements of adherence. They are encountered in both microorganisms and on the surface of tissues, in all living beings. They will therefore be studied as a whole. Some correspond to veritable anchorage devices, with precise structures, whereas others are reduced to gels, mucuses, soluble substances, but are nevertheless active. Besides these microstructures with the more or less viscous, gellified, or openly liquid coats, room should also be reserved for the ionic zones genuine "aura" which also take an active part in the mechanisms of adherence.

3.1. Adsorption — Ionic Zone

This very classic zone shows adsorption, the first stage of adherence. All the interfaces, whatever their natures, have properties of attracting substances dispersed or dissolved nearby. That is how the phenomenon of adsorption, studied by physicists for years, was defined. Everyone knows that solid surfaces adsorb gases just as well as other substances. Numerous applications have been drawn from that knowledge. They have become commonplace, as is the case for the properties of kaolin, clays, fine sands, glass powders, and more modern materials such as ion exchangers. It is also well known that bacteria are carriers of electric charges, in certain medium conditions, and depending on these charges, of Van der Waals-type forces of attraction; electrostatic forces intervene during the first phase of the film leading to irreversible adherence. Organic or mineral-suspended matter obey these laws of physics concerning adsorption. It was accepted that the attachment of microbes on interfaces was limited to these physical interventions alone. The energetic relations between ordinary particles are undoubtedly governed by a certain number of factors among which the importance of respective distances, electric charges, dimension, and the nature of these particles themselves can be noted. The observation of the microbial world presents small-sized corpuscles, submitted to the hazards of probabilities, to the laws that Boltzmann formulated taking into account the ambient temperature.

Polar actions intervening in the behavior of bacteria were the subject of an in-depth study in 1993, by J. Maddock et al.[14] The studies were based on strains of *Listeria monocytogenes, Caulobacter crescentus,* and *Escherichia coli.* There is indeed a specific polar location in relation to the surface proteins, including receptive chemocomplexes. The authors believe that bacteria are capable of "choosing" between a new pole and the former pole and determining their behavior: a teleopathic thesis already supported by other authors as previously mentioned. Deneke et al.[15,16] (1975) provided a detailed analysis of it. Lips and Jessup[17] (1979) distinguished conservative forces resulting from the relative proximity of the elements present. Such is the case for electrostatic and electrodynamic forces. Other attractions respond to the movements of suspended matter in a liquid or gaseous medium. Aerodynamic and hydrodynamic forces are put forward as examples. Lastly, the authors admit the existence of so-called "outside" forces, represented by electric and magnetic fields. They are qualified as "gravitational".

Everyone admits that it is difficult, if not impossible, to solve the problems of adherence with the help of physics alone, especially when it is a matter of living beings that adapt poorly to such reductionism. It is naturally agreed that these laws account for a certain number of actions between inert substances of known composition. Nevertheless, the knowledge pertaining to microbial chemical structures and molecular architectures, including those of viruses, advances, certain aspects of physics, and its teachings, provide useful arguments. For example, reference is made to the specificity of the forces related to the structure of the bacterial walls, thanks to which it is possible to account for preferential adherences such as the *Staphylococcus*/protein system. The same is true for the attachment of viruses on clays, glass powders, and sands depending on the pH conditions.

Numerous practical applications have been drawn from these observations and experimental data, notably for the determination and isolation of viruses present in water or mollusks (J. Nestor, J. Brisou, F. Denis, H. Shuval, et. al.).[18-20] When microorganisms rub shoulders with inert surfaces, the first stage of adhesion is obviously guaranteed by these commonplace physicochemical factors. The small size of microbes makes it possible to assimilate them to living "colloids", as suggested by Winogradsky[21] in 1925. Due to this comparison, it became customary to call on these varied attractive electromagnetic forces and to apply them to the behavior of the infinitely small in nature. Ionic forces are obviously related to the presence of groups COO–, PO3–, NH^{3+}, etc. The analysis of these electrostatic relations contracted between the bacteria themselves, or between varied interfaces, was presented in 1941 and in 1948 by Derjaguin and Landau[22] on the one hand and Verney and Overbeek[23] on the other; hence the initials DLVO adopted for referring to the theses presented by these authors.

Nevertheless, the range of the forces developed remains limited when one considers the difficulty of imprisoning the living world in physical and mathematical models. No one questions the role of ionic forces — and we will have the opportunity of recalling this supported by examples — but it is reasonable to be aware of the limits of these roles.

Following is the relation summarizing part of these activities: VT = VA + VR, in which VT corresponds to potential energy, VA to the force of attraction, and VR to that of repulsion. Many authors have attempted to make this expression coincide with the behavior of bacteria between each other or vis-a-vis varied surfaces. The majority of them side with previously formulated conclusions and no one questions the fact that all the particles carrying electric charges are subject to the general laws mentioned earlier.

Mobility is a function of the viscosity, of the fluidity of the liquids containing this suspended matter, its ionic composition, and the nature of the interfaces present. It is known, for example, that for a pH of 7.0, the mobility is the same for all bacteria. It is naturally only a matter of Brownian mobility. Mobility due to the presence of locomotor cilia is excluded from this discussion. Experience teaches, for example, that the alteration of the hyaluronic coat of *Staphylococcus aureus* causes a substantial decrease in Brownian agitation. This drop is attributed to the loss of COO·(Hill et al., 1963).[24] The behavior of this same bacteria was observed at different pH levels and its Brownian agility reaches a maximum between values ranging from three to four, which indicates the presence of ionic groups with low pH. A brief treatment at 0° using metaperiodate for a pH of seven completely modifies the behavior of the microbial cells. Rogers[25] (1963) and Garrett[26] (1965) suggested that the periodate oxidizes the teichoic acid and that this modification of a chemical order explains the results observed. This acid will be studied in the next chapter with the chemical agents of adherence.

Any modification of the chemical or even stereochemical structures is likely to cause a different behavior of the microbes concerned. The sensitivity of certain bacteria in the ionic conditions of the immediate environment caught the interest of other researchers: Szmelcman (1978), for example, noted the behavior of *E. coli* confronted with ionic variations. He supported a debatable teleological thesis suggesting the existence of a veritable sensorial system organized in these primitive beings, and the possibility of a transfer of information ranging from cellular receptors to the flagellary "motor". Modifications of membrane potential were demonstrated (Szmelcman and Adler, 1976). Qualitative and quantitative variations of the extra- and intracellular ionic state were confirmed in the presence and absence alike of ionophores such as valinomycin and nigericin. The authors noted a permeabilization to protons. Such experience confirms the variations of bacteria mobility depending on the ionic components of the microenvironments in which they are called upon to live and evolve.

In reality, these facts do not offer anything surprising, and it should be noted that this ionic "aura" is sometimes favorable and sometimes harmful to adsorption within the boundaries of a microenvironment which should not be neglected. Such a comment is aimed at the abuse committed by certain authors who, base their statements on the unique data obtained *in vitro*, stoop to generalizations which, as far as nature and the ecology are concerned, remain questionable. Clearly, understood epistemology teaches that the observation of limited phenomenon, taking place in the laboratory, does not authorize such generalizations. In nature, only "snapshots" can be taken in extremely complex conditions. However, in the laboratory, we are free to observe in conditions of exact time, at regulated temperatures, in chosen environments of microorganisms selected, imprisoned, tamed, and subjected to our every will.

Nevertheless, it should be recalled that the concept of "chemotactism" described at the end of the nineteenth century by Engelmann, applied by Pfeiffer and Stahl, cannot be overlooked. It plays a certain role in the mechanisms of adherence during the first stages of the process. The old treaties describe methods of varying accuracy which were successfully applied for the isolation of mobile bacteria and attracted by certain substances. The germs equipped with locomotor cilia lent themselves particularly well to these techniques that bacteriologists used in the past. Recently, Lengeler[27] devoted a well-documented article (1990) to the "swimming movement" of bacteria. The operation and structure of the bacterial flagella are well known. The *Salmonella typhimurium* flagelline, for example, is rich in *N*-methyllysine residue which also exists in the actin in muscles. The molecular weight of these flagellines can reach 40,000 Da. Hinkle and MacCarty (1977) emphasized the role of ATP in the triggering of the helicoidal movement of these flagella. The fact that the bacteria show affinities or repulsions for one substance or another must intervene in adherence and examples of this will be given, notably in the

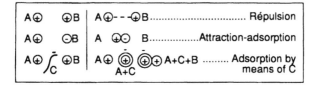

Figure 2. Ionic relations between adhesins.

chemical recognitions that are studied later. However, it is essential, before tackling this subject, to review a few common concepts pertaining to attraction forces (see Figure 2):

a. *Ionic bonds:* They correspond to basic unions between ions of opposite charges. They result from the loss or gain of electrons.
b. *Covalent bonds:* They are made when two atoms pool their external electrons.
c. *Metal bonds:* They characterize the metals freed peripheral electrons which move and act as so-called "conduction" electrons.
d. *van der Waals forces:* They achieve the uniting of atoms or neutral molecules close to each other. They are weak and fragile bonds.
e. *Hydrogen bonds:* A hydrogen atom ensures the union, also fragile, between two neighboring molecules. Biochemistry provides many examples of them.

3.2. The Roles of Polycations

Polycations have mainly been studied in adherence mechanisms of vegetal viruses. Poly-L-ornithine was the material chosen by Watts et al.[28] in 1981. The virus and polycation were preincubated for 10 minutes before inoculation in the plant. The virus was that of the tobacco mosaic. It was noted that this treatment promotes the penetration in the protoplasts of the plant. A diagram inspired by the authors makes it easier to understand the modalities of this process. The formula of this amino acid: –d-diamino valerianic acid is recalled below. It is known that ornithine does not participate in the structure of proteins, but that it plays an important role in the metabolism of urea. The virus of the tobacco mosaic is negatively charged. The polycation is positive; the plasmalemma is also, like the virus, negatively charged. It therefore repels the virus. The polycation changes the charge of the virus either by neutralizing it or by reversing the sign. From then on, it will be attracted by the plasmolemma. The authors were able to count the optimal proportions of poly-L-ornithine and of the virus likely to lead to maximal effectiveness, therefore to the optimal level of infectivity of the virus. It appears to be 1 µg/ml of polycation for 10 µg/ml of virus. The same phenomenon was observed in Mycoplasma and other bacteria (*Pseudomonas, Klebsiella, Nesseriaceae,* etc.). (See Figure 3.)

The general formula of amino acids - excellent polycations - can be summarized as follows:

$$CH_3 - (CH_2)_n - \overset{\overset{\displaystyle H}{|}}{\underset{\underset{\displaystyle NH_2}{|}}{C}} - \overset{\overset{\displaystyle O}{\parallel}}{C} - OH$$

ornithine, experimented here, is presented (a – d – diaminie)

$$NH_2 - CH_2 - CH_2 - CH_2 - \overset{\overset{\displaystyle H}{|}}{\underset{\underset{\displaystyle NH_2}{|}}{C}} - \overset{\overset{\displaystyle O}{\parallel}}{C} - OH$$

symbolized by this schema

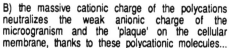

for a total of 12 cations and one anion

A) a negatively charged microorganism will be repulsed by cells of the same charge

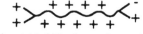

B) the massive cationic charge of the polycations neutralizes the weak anionic charge of the microogranism and the 'plaque' on the cellular membrane, thanks to these polycationic molecules...

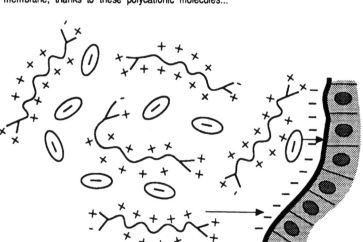

Virus, chlamydia, for example, are thus transported, adsorbed and may possibly penetrate the eukaryotic cells.

Figure 3.   Role of polycations.

3.3. Adhesion — Second Reversible Stage

This second stage is marked by the recognition of elements that may be likely to promote adherence. The stereospecificity between certain patterns of host organs called receptors and the bacterial structures characterize this approach. There is a similarity with an enzymatic reaction where a substrate and an enzyme meet and unite specifically to give birth to a new product. The very classic relation between the two partners is formulated as follows: E-S → the ES complex, which after the E + S reaction produces P.

As far as adhesion is concerned, the two facts will be: the cellular receptor R and the bacteria B. During the encounter, there will therefore be the complex RB, a transitory, reversible encounter that will stabilize in function with the structural complementarities and especially the aptitude of the bacteria to synthesize the anchorage organs in charge of ensuring the solidity of the following stage and its irreversibility. The reaction, in accordance with the laws of Michaelis, can therefore be summarized in the following way: in which R represents the receptors and B the bacteria, during the encounter, the instable complex RB is formed:

$$R \text{ meets } B = R \Leftrightarrow \text{ reversible } B, \text{ Adhesion}$$

If the bacteria already have anchorage organs or are capable of synthesizing them, the final, irreversible adherence will be guaranteed.

$$R + B \Rightarrow \text{ Adherence}$$

Here again we can, by analogy, refer to enzymology by taking into account affinity constants and the ratio of mass between the receptors and the number of bacteria, namely R/B, which boils down to considering the ratios between the concentration of a substrate and the affinity constant of the enzyme. A very simplified diagram summarizes the process (Figure 4). In the present case, the product formed will be the number of successful adherences. Here the ratio R/B will only be 0:5.

3.4. Irreversible Adherence — Adhesins

3.4.1. Microcapsules and Capsules

Here, we are in the presence of genuine organs of cellular coating that electron microscopy reveals to us. Microcapsules were described in 1958 by Wilkinson.[29] They were assimilated to S (smooth) antigens of Gram-negative bacteria and related substances, such as ectoproteins different from the lectins that will be studied shortly. Those of the hemolytic streptococci known by the name of "M" proteins were included in this category. They are now considered as authentic lectins. The "O" antigens, glucido-peptido-lipidic antigens, do not fall into this category. The real capsules are, however, well defined and widely

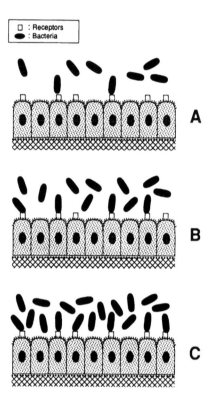

Figure 4. Application of Michaelis' law to the adhesion of bacteria on an epithelium.

described in the classic treaties. It is therefore unnecessary to spend time on them. Everyone knows they provide the protection of bacterial cells. Their polysaccharide nature is recognized. The components are frequently rather close to the parietal elements, notably when they contain uronic acids. The capsular polysaccharide of *Bacillus megaterium* is, for example, immunologically identical to the parietal that contains glucosamine and muramic acid. A phosphate ribitol was evidenced in the capsular polysaccharide of *Pneumococcus* VI, formed of galactosyl-l-glucosyl-rhamnosyl-ribitol units that unite diester bridges flanked by a ribitol at one neighboring end with galactose (Reters, Heidelberg, et al., 1961).[30]

While studying the capsule of *S. aureus*, Wiley and Wannacott (1962) revealed the presence of the isomer D of glutamic acid, lysine, alanine, and glycine found in the composition of the rigid wall. This capsule also contains glycerophosphate. Glucose, glucuronic acids, galactose, and fructose are commonly found in the composition of capsular structures and play a capital role in the mechanisms of adherence. Their presence directs the possible action of

  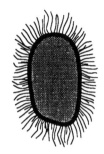

| Leusonostoc mesenteroid dextran covering | Leusonostoc mesenteroid dextran covering after culture in the presence of saccharose | Escherichia coli rich armor of varied fibers-fimbriae |

Figure 5. *Leusonostoc mesenteroid* **and** *Escherichia coli.*

specific enzymes and will be discussed later. Gums and waxes can, moreover, "coat" certain bacteria (Mycobacteria, among others – see Part I, Chapter 1). It is known that these capsules play an important role in the mechanisms of pathogenicity, but it is admittedly not exclusive. The same is true for antigenicity and also for the immunogenic power. Nevertheless, it will be shown shortly that the role of the pili now seems to occupy the forefront.

It should be specified that certain capsular polysaccharides themselves cover fimbriae, which does not simplify the problem (Dazzo et al.,[31] 1986). (See Figures 5–8.)

3.4.2. Microtubules

These microtubular systems only concern epitheliums with respect to adherence. They were evidenced on the intestinal epithelium of rats and studied closely by Bader and Monet[32] in 1978. It is known that they are formations that come to life during cellular division. They come from well-known and morphologically individualized cellular zones. Their distribution undergoes profound change during cell division. Residues are often observed which, after division, still serve as links between the two daughter cells; hence, the suggestion of a possible intervention in the adherence process. Such structures were observed on the vesicular epithelium, on kidneys in various mammals, and on branchial tissues in fish. They are even visible in many protozoa, in short, in all the cells subject to division by mitosis, which excludes prokaryotes. These microtubules participate above all in ionic exchanges. Substances such as alkaloids extracted from periwinkles, for example, provoke disorders opposing themselves to the transfer of Ca_2, Na^+, PO_4 and H_2O. These formations are also the theater of phosphorylations. The value of surface polysaccharides in cells during biosynthesis is therefore understood.

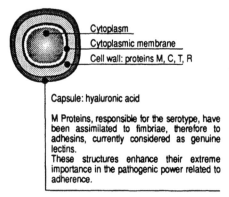

Figure 6. Envelope of a streptococci.

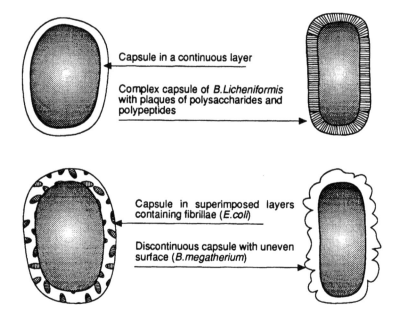

Figure 7. Types of bacterial capsules.

3.4.3. Lectins

3.4.3.1. General

The most striking mechanisms of adherence are now recognized as being closely linked to a vast group of glycoproteins known by the name of lectins. Discovered in 1888 by Stillmark,[33] at the beginning they only interested a few specialists whose attention was caught by surprising reactions of agglutination

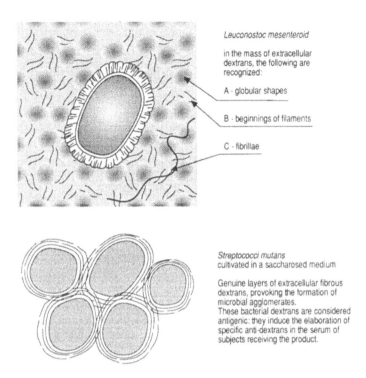

Leuconostoc mesenteroid

in the mass of extracellular dextrans, the following are recognized:

A - globular shapes

B - beginnings of filaments

C - fibrillae

Streptococci mutans
cultivated in a saccharosed medium

Genuine layers of extracellular fibrous dextrans, provoking the formation of microbial agglomerates.
These bacterial dextrans are considered antigenic: they induce the elaboration of specific anti-dextrans in the serum of subjects receiving the product.

Figure 8.   *Leuconostoc mesenteroid* **and** *Streptococci mutans.*

of animal or human erythrocytes in the presence of vegetal seed extracts. Easily observed, not requiring any costly material, the phenomenon of hemagglutination occupied a growing number of researchers. All kinds of seeds, then plants, were ground to study the behavior of erythrocytes in the presence of the extracts obtained. All these substances were grouped together under the name of "Phytoagglutinins" or "Phytohemagglutinins". In 1908, Guyot[34] showed that certain *E. coli* strains can, like the vegetal extracts, provoke agglutination of red cells. The field of agglutinins was widening. Much later, Kauffmann[35] resumed these studies with the isolated strains of appendicular peritonitis. This is still a current topic of research, as will be demonstrated in a moment. Nevertheless, it was not until 1954 that Boyd and Shapleigh[36] gave these singular and more or less mysterious substances the name of lectins, judiciously inspired from the latin *lego, legere* meaning "recognize, choose, sort". These complexes do indeed have the essential property of recognizing and retaining certain specific molecular patterns and attaching themselves to them. In nature, they are always carbohydrate patterns. In 1949, Boyd and Reguera[37] noted that the extracts of *Phaseolatus lunatus* (lima bean) agglutinated the human erythrocytes of group A. Renkonen, a year earlier, had experimented on 57 vegetal extracts belonging to 26 types. Such investigations were actively continued, showing that lectins are in fact extremely widespread in living beings since they are found in bacteria and plants, animals and humans.

As early as 1891, Erlich[38] succeeded in immunizing mice against the toxin of ricin and the jequirity bean (Abrin of *Abrus precatorius*) by injecting small repeated doses of vegetal extracts into animals. For the author, this was the opportunity to suggest the first method of in vitro titration of antibodies, through the action of the antiserum on the vegetal toxin (1897).[39] The reagent making it possible to follow the phenomenon is simple since it is a matter of washed erythrocytes, normally agglutinated by the extracts. The antiserum inhibited the reaction. In 1901, Landsteiner[40] used abrin for similar research. Phytoagglutinins quite naturally fall into the field of immunology. They have been solidly rooted there for over 80 years.

These observations allow indispensable immunizing properties to be attributed to certain lectins and leave the door open to a short excursion into the anecdotal history about "Mithridatization", the famous Mithridates VI Eupator who lived around 63 B.C. Convinced that his entourage was trying to poison him, he regularly absorbed small quantities of vegetal extracts considered as toxic. The plan succeeded since the emperor resisted the murder attempt concocted by his son, Pharnace. That is a simple, entirely personal hypothesis for explaining a remarkable episode, or legend.

To complete this excursion into the past, it should be pointed out that in 1884, thus before the discovery of hemagglutination, Bruylants and Venneman,[41] Belgian biologists, became interested in the effects of jequirity bean extracts on animals and humans. These authors compared the extracts to an enzyme. They showed the inflammatory properties on conjunctiva and the toxicity in rabbits and frogs. Without noting the *in vitro* hemagglutinating effects, they nevertheless pointed out "substantial vascular swelling" of the interdigital capillaries in frogs after administration of jequirytin. This event was in all likelihood due to a massive *in vivo* hemagglutination under the action of the toxin.

Bacteriologists, cancerologists, hematologists, geneticists, etc., call on lectins more and more frequently. They study their outstanding properties and make them genuine work tools. A review of the bacteriological applications was made in 1985 by C. Dieuaide.[42] More than a hundred lectins are now marketed and numerous works are devoted to them.

Lectins are, therefore, glycoproteins gifted with a strong affinity for certain saccharide sequences, or for isolated sugars, such as mannose. In general, they have two binding sites. Whereas some show a preference for a non-reducer terminal sugar, others can bind with internal sequences taking part in the structure of macromolecules. Nevertheless, it is accepted that all lectins express a certain degree of discrimination between the carbohydrate sequences. Many of them have an elective sensitivity directed towards a single type of carbohydrate, but this property does not exclude other less privileged potentialities. Such a concept widens their field of activity and explains the scope of applications that are drawn from them. A lectin makes it possible, for example, to distinguish in the heart of a macromolecule, certain saccharide sequences, or to identify non-reducer terminal residues, as shown by Cummings and Kornfeld[43] in 1982 and then by Crane and his colleagues[44] in 1982.

The affinity that these substances feel for simple carbohydrates, oligosaccharides, does not necessarily faithfully reflect the exact image of the materials present in a macromolecule. These weaknesses in the complexity of steric configurations of spatial structures lead to the inter-reaction of lectin and several saccharide chains instead of limiting their action to one structure. Everything is related to the potential, to the power of affinity between the molecular patterns and to stereochemistry. Shortly, it will be noted that molecules must be considered (included) depending on their very complex, battered, polylobed, spatial construction, rich in cavities of varying depths, such as those seen in a variety of macromolecules (enzymes). The observations made over the past years show that lectins react with the oligosaccharides or glycoproteins of cellular interfaces. Consequently, any oligosaccharide or monosaccharide having a specific affinity can inhibit or reverse a normally predictable reaction.

Before proceeding any further in the study of these complexes, it appears essential to specify that two large groups can be distinguished: (1) Attached lectins called structural lectins, and (2) Soluble or viscous lectins. These groups can all be coupled by covalent bonds with ferritin based on techniques developed by Kishida et al.[45] in 1975.

3.4.3.2. Pili I or fimbriae structural lectins

These complexes correspond for the most part to those "cellular receptors" described by the classics. In the strictest sense of the term, they belong to surface structures. They are encountered either in connection with glucoconjugated coats, or in the form of microvesicles. In bacteria, they are mixed with type I fimbriae, also called "Common Pili".

The membrane lectins of hepatocytes fix, in many mammals, galactose to form asialoglycoprotein complexes where galactose terminal residues have been identified. These lectins certainly play a role in the metabolism of sialic acid and glycoproteins in general. Other receptors contract bonds with substances such as mannose-6-phosphate, N-acetylglucosamine, mannose, etc. Specialists obviously marked these lectins with ferritin, fluorescein, or radioelements in order to locate them with as much accuracy as possible and to follow their course and attachment on other cells.

Carcinologists also took advantage of this opportunity to experiment with them on neoplasic cells with interesting results: Inbar and Sachs (1969), Lis et al.[46] (1973), and Nicolson[47] (1974) presented their observations. However, the analysis of these studies will not be dealt with here so as to remain within the field of microbiology.

Fimbriae, the Latin translation of "filaments", represents genuine pilous systems and lead straight into the field of adhesins which can be qualified as "major". Thanks to electron microscopy, these organella were revealed to Duguid et al.[48,49] in 1955. These extremely thin filaments cover the entire surface of certain microbes; enterobacteria adorn themselves with these filaments willingly. A few commonplace examples can be given: *E. coli* and *Proteus* but also *Pseudomonas, Agrobacterium, Flexibacter, Neisseria, Rhizobium*, etc., and even chlorophylian germs, streptococci, corynebacteria, etc. These fimbriae are composed of fibers ranging from 48 to 85 Å in diameter for an average length of 1 µm. Duguid[50] divided them into four groups.

The representatives of group I (common pili) concern bacteria, lower fungi, plants and certain animal tissues including erythrocytes. The agglutinating effect of certain sugars such as D-mannose, CH_3-alpha-D-mannosid is exercised at this level, with respect to a variety of germs that will be examined with a few details. As the position of C_1 in the mannose molecule is not indifferent, it provides an example of a stereochemical intervention in the specificity of adherence. The colonies, masses in stars observed in certain microbial cultures as the veils on liquid mediums, constitute many expressions of these agglutinates due to the entanglement of these pilous surface microsystems. Aquatic bacteria, such as cytophages (the so-called "Slipping/Gliding" bacteria, represented by *Flexibacter)* and *F. polymorphus,* are agglomerated depending on the nature and layout of their fimbriae. According to work by Wistreich and Baker, the *Neisseria* covered by fimbriae provoke hemagglutinations usually resistant to the inhibitor action of D-mannose. The relative frequency of mutants deprived of this ectopilosity gives an idea of the extreme variety of the adherence mechanisms the progess of which is practically impossible to predict. The analysis of these fibers rather easy to detach from the germs, enabled chemists to estimate the proportion in amino acids at 9% foundations of a specific protein qualified as "Pilin", a substance recognized as being antigenic. The specific weight is about 16,000 Da. All the works reproduce excellent images of these fimbriae observed using electron microscopy.

A considerable number of these organisms have been described over the past 20 years. Sialic acid appears to be one of the essential components. For example, sialic acid is found in the fimbriae of many strains of *E. coli, Streptococci, Neisseria, Pseudomonas*, etc. It is known that these bacteria colonize the mucosa of humans and animals. The lectins located at the surface of these fimbriae facilitate colonization thanks to the affinity of terminal sialic residues with the mucins coating these mucous surfaces. However, the carbohydrate structure only provides part of the explanation of this adherence.

The *O*-acetylation of sialic acids may play an important role according to Schauer (1982). The work by Duguid et al.[48,49] clearly established that the "mannose sensitive" character of the hemagglutination was related to the nature of the fimbriae. It was believed that these organella gave the adhesive properties of the bacteria to all the cells: erythrocytes, vegetal epitheliums, etc. However, now we know that other fimbriae are insensitive to mannose. The presence of different types of pili led specialists to reserve the name type I

"Pili" for these fimbriae. In sum, they are the most striking elements, and sometimes 100 to 400 of them are counted on a *E. coli*-size bacterial cell.

A. Brinton[51] (1967) gave specifications on the physicochemistry of these structural surface elements. They can be detached from their cells and purified. The pilin of *E. coli*, isolated and prepared by Brinton[51] himself, contains a high proportion of hydrophobic amino acids. Their protein sequences are known.

It can be concluded that the type I fimbriae are slightly lipophilic and contribute to the formation of aggregates, masses, and angular packs, the specialized structures of which provide excellent, demonstrative images.

3.4.3.3. A few properties of the fimbriae

Ofek et al.[52] (1978) observed the behavior of *E. coli* in the presence of buccal epithelium and showed the sensitivity of certain strains to mannose. They attributed this singularity to the presence of mediator lectins present at the surface of the germs. It was suggested that the germs unite with the mannose residues that the epithelial interfaces offer them. This thesis was supported by facts making it possible to note that adherence was, in certain conditions, inhibited by the treatment of the cells with sodium metaperiodate. This product oxidizes the carbohydrate residues present at the surface of the cells. The authors obtained analogous results, after treatment using concanavalin A, of which the affinity for mannose is known. On the other hand, pea and wheat germ lectins, which have no tendency to unite with mannose, remain indifferent to the cells.

These facts can be taken as an example of what others report in chemotactism which falls within the framework of stereochemistry and molecular recognitions. These observations agree with what is known about the agglutination of "mannose sensitive" bacteria and the cells of *Saccharomyces cerevisiae, Candida albicans* the surface structures of which contain mannans. Here again, oxidation using periodate inhibits the agglutination of *E. coli*. There is therefore an unquestionable correlation between the affinity of bacteria with mannans and the facility with which they adhere to animal cells (Z. Bar-Shavit et al.[53] 1980). From these observations, it was deduced that type I fimbriae of *E. coli* correspond to a well-defined lectin. It appears that the attachment sites providing these coating organella the sensitivity to mannose (MS = "mannose sensitive") are spread over the entire surface, if credit is given to the conclusions formulated by Sharon et al. in 1986. (See Figure 9.)

Without going into the technical details of purification of the fimbriae (Brinton, Salit et al., Korhonen et al., Eshdat,[51,55,60] etc.), it should simply be recalled that the type I fimbriae cover a large number of Enterobacteria strains. The composition in amino acids and the sequences in a few strains are partially known. To confirm this observation, a few examples borrowed from Sharon and Ofek[54] (1986) follow (residues per mol).

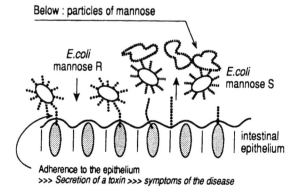

Figure 9. Example of trapping by specific molecules.

The lectin of *E. coli* K99 is a homopolymer of subunits, the molecular weight of which is close to 18,500 Da. The strain agglutinates erythrocytes in horses, pigs, and humans. Its action is inhibited by the ganglioside GM_2 and by a prior treatment of erythrocytes with sialidase, or with the neuraminic *N*-acetyl acid. The glycophorin of the erythrocytes behaves as a specific receptor of this K99 strain.

The antigenic properties of the fimbriae are widely confirmed. The isolated elements of *E. coli* contain a specific antigen; the same is true for the fimbriae of *Shigella flexneri,* which contain a small fraction shared with the two other strains of *E. coli.* Count of the amino acids contained in the fimbriae was as follows. Examination of 79 serotypes of *Salmonella* and a few strains of *Arizonae* and *Citrobacter* revealed the existence on the fimbriae of antigens shared by all. However, there is no relation between the *Shigella-Arizona* group on the one hand and the *Shigella-coli* group on the other. The study of 4 serotypes of *E. coli* showed that the strains have antigenic determinants in common on the fimbriae, but that each one also has additional specific patterns.

A diagram inspired by Brinton[55] (1965) represents a type I fimbriae decorated with bond sites identified thanks to the monoclonal antibodies by Abraham et al.[56] in 1983. These authors used three anti-fimbriae monoclonal antibodies of *E. coli.* The AA_8 and CG_1 antibodies do not attach themselves to the surface of the fimbriae, nor on the bacterial cells, but they react with a subunit of monomers and with dimers of fimbrial origin. However, the antibody, called CD_3, reacts solely with hexamers weighing about 102,000 Da or with oligomers of equal or higher weight. The results published in this way lead to the conclusion that certain fimbrial epitopes are dependent on determinants with a quaternary structure. Only the CD_3 antibody is opposed to the adherence of the bacteria with epithelial cells or with guinea pig erythrocytes. The important concept of spatial structure arises here. The affinity of fimbrial lectins for certain sugars is now well established thanks to research devoted to the inhibitor effects of adherence by these carbohydrates, whether present in the form of monosaccharides or oligosaccharides.

In one study, *E. coli*-yeast served as a model for Firon et al.[57,58] in 1983–84. This cellular complex forms an easily observed aggregate. The authors experimented with a strain of *E. coli* 346, the fimbriae of which were isolated. They noted the effects of carbohydrates or polycarbohydrates likely to inhibit agglutination. The structure of these inhibitors suggests that the most favorable to the unions correspond, here again, to the dimensions and shape of a trisaccharide. The existence of depressions or real pockets at the surface of the lectins is put forward. This recalls what is observed in enzymes and antigens. That is how the definition of epitopes and paratopes is reached, or the old comparison between the key and the lock.

In the case of *E. coli* fimbriae, it is estimated that there are at least 3 secondary sites, each one of which is capable of latching onto a monosaccharide residue. The high inhibitor power of *p*-nitrophenyl-alpha-mannoside is attributed to the presence of a hydrophobic pole next to one of the secondary sites (Sharon et al.,[59] 1986).

Other strains have naturally been examined; the behavior of *Kl. pneumoniae* is very similar to that of *E. coli* 346, whereas the salmonella act differently. With six experimented strains, *p*-nitrophenyl mannosid only shows very modest inhibitor effects. It is estimated that the receptor sites of *Salmonella* are smaller in size than the *E. coli* and *Klebsiella*, and that they do not have hydrophobic zones. These observations show that the sensitivity of the fimbriae to the carbohydrates often remains temperamental, varying with the strains and sugars. These properties are, without a doubt, governed by genetics as will be shown in a moment. Products such as *p*-nitrophenyl mannosid, trisaccharides of alpha mannosid, manna-3-Man-beta 4 Glc-Nac, substantial inhibitors of *E. coli* agglutination, act very differently in the presence of *Salmonella, Klebsiella, Serratia marcesens, Enterobacter clocae* or *Enterobacter agglomerans*, for example.

On the other hand, lectins were discovered on the flagella themselves which does not simplify the problem. They were evidenced on a strain of *E. coli* 7343 which is a protein clearly different from the fimbrial lectins. Likewise, flagellar lectins of *S. marcescens* 8347 are known, evidenced by Eshdat et al.[60] in 1980 and 1981. Such substances should no longer be foreign to chemotactic demonstrations. They also constitute biochemical signals. The work by the authors mentioned confirms the presence of other structural lectins which are present in the form of small vesicles located on the membranes. These formations may be pilin-based monomer fimbriae inserted outside the membrane without polymerization, or even protein subunits different from pilin.

In short, it stands out that glycoprotein structural lectins (the usual aspect of which merges with the fimbriae) exist, but that these organella are not alone. Other lectins have been identified on the pili, flagella, and membranes. Although they are different from each other on several points, they share the property of recognizing simple or complex carbohydrate patterns. Certain authors plan on interpreting these substances as evolutive "phases" of close complexes which, depending on their position in the cellular architecture, acquire their originality. This suggestion makes it possible to understand this diversity of bacterial behavior. These are indeed evolutive states of substances devoted to similar functions of recognition and bonding. For example, Le Minor et al.[61] (1973) have observed the reaction between the lectin of concanavalin A and the O antigen of a few enterobacteriaceae.

3.4.3.4. Soluble lectins

These substances lead us to a completely different field due to their moving, mobile, unpredictable character. Nevertheless, we will stay within the widened framework of glycoproteins capable of recognizing, in accordance with the definition, carbohydrate patterns. The course of these lectins is obviously becoming more difficult to follow; they do not always let themselves be observed carefully. Research made since the last century (1888) on seed, plant, and tissue extracts from every possible animal (invertebrates, fish, a variety of animals, mammals, including man) suggests that these glycoproteins are developed inside the cells and then released in the immediate environment. They then contract unions with complementary glucoconjugates.

At the time of their discovery and due to the results obtained, it was accepted that they constitute a new class of extracellular proteins raising many questions for specialists, who were puzzled by the often polyvalent character of these compounds. They organize themselves in macromolecular masses that cover, surround, and literally wrap the cells. All living beings, from bacteria to humans, produce them. They are encountered, without exception, at all the levels of evolution. Confusion with other substances should be avoided, the lectin-like also being of cellular origin, as specified by Goldstein et al.[62] in 1980. Most important, it should be noted that these soluble lectins are above all gelifying, precipitating, hemagglutinating substances expressing themselves as soon as they encounter carbohydrate patterns in relation to which they have specific affinities.

They are sometimes monomeric and sometimes polymeric. Fibronectin or laminin, for example, is combined with heparin and hemagglutinates. The elaboration of some of these lectins begins from the first stages of embryonic life, reaching a maximal level of activity at the producer organism's age of adulthood. During studies devoted to the relations between lectins and ontogenesis, it was recognized that they are really products of cellular secretion, the beginning of which coincides with the formation of the blastula, therefore very early. In cellular cultures, conditions of fasting induce the lectin secretion. The elements agglomerate, form blocks of varying compactness, and organize themselves in filamentous masses. Such observations were made on cultures of lower fungi; *Dictyostelium discoideum*, for example, delivered a substance that was purified called "discoidine". Two variants, I and II, are distinguished, the structures of which are now known.

All in all, soluble lectins are presented in the chemical form of glycoproteins directly secreted by the cells. They can stagnate at the surface of producer tissues or spread out in the immediate environment. Their agglutinating properties, the preferential bonds they contract with carbohydrate patterns, and their specificity of action all make them participate in the organization of tissues in higher beings, and in forming aggregates in microorganisms. They play an unquestionable role in the adhesions, adherence, and attachment of microbes on tissues and facilitate the gregarism of unicellular beings.

In higher beings, they participate in the constitution of gels, mucus that line the mucosa, and a variety of epitheliums. They unite with various glucoconjugates and participate in numerous reactions on surfaces of which the wealth in glycoproteins, glycolipids and polysaccharides is widely asserted. A lectin can only bind after the construction of multiple "bridges" between the opposed cells (L. Lis and N. Sharon,[63] 1986). Erythrocytes submitted to a high concentration of lectin may remain indifferent. It is therefore estimated that other factors intervene in the process, notably the physical and chemical conditions of the microenvironment, and the state of the receptor cells themselves. Thus, when these erythrocytes were submitted to the action of enzymes such as glucosidases, trypsin, sialidase, and the soy bean lectin that normally causes agglutinations, they had a totally unexpected behavior. The agglutinations were sometimes accelerated for one and, for the same dose of lectin, were sometimes inhibited.

Pusztal et al.[64] (1993) reported that the lectin of the ordinary bean (*Phaseolus vulgaris*) introduced in the daily diet of rats promotes proliferation — invasion of the rat's small intestine by mannose-sensitive *E. coli*. This lectin is curiously inhibited by an extract from a "snowdrop" bulb (*Galanthus nivalis*) added to the food ration. The invasion of the bacteria is pratically stopped. The authors suggest that this plant extract could be used for checking the invasive power of the bacteria on the small intestine. They nevertheless emphasize that this extract does not neutralize the pathogenic power of the *E. coli* at issue. These observations, combined with many others published in the literature, would be favorable to supporters of certain phytotherapies and phytoprevention.

We will have the opportunity to return to this subject in the chapter devoted to medical applications. It is easy to understand the interest that industry has for these substances, which are universally widespread in the world of the living.

3.4.3.5. Classification

Diversity calls for putting into order, classifying in categories. The models put forward by several authors are naturally not all unanimously accepted. We will stick with the basic and simple.

Doyle et al.[65] take into account the superficial components likely to contract bonds with the lectins. Other authors prefer a classification based on the origin of the substances. Thus, ricin, abrin, limulin, and tridacin are recognized. For the most part, trade adopts this trend due to its convenience. The origin of lectins is known immediately, but its original properties are unknown. That is undoubtedly why Prokop et al.[66] suggested basing them on the criteria of agglutinated erythrocytes. Mäkelä,[67] going even further, bases his classification on stereochemical considerations. He distinguishes four groups of substances that differentiate the position of the carbons in C_3 and C_4 of pyranose. Based on the same principle, Rupe limits the system to only three groups. It would be vain to side with one or the other system. A chart from the work by Doyle et al.[65] is provided as an example of these classifications.

In 1984, Gallacher[68,69] leaned toward a more original idea. He imagined the breakdown of these recognition complexes in two large classes, based on their affinities for sugars. Class I brings together lectins attracted by monosaccharides participating in the structure of a complex. In these conditions, the presence of this free monosaccharide in the medium opposes itself to the union, by competition and by trapping the lectin. In this class I, including what he calls "Exolectins", Gallacher differentiates: (1) sub-class Ia unites mandatory exolectins, the affinity of which guides them towards the sugars located at the end of a macromolecular chain; (2) sub-class Ib brings together less demanding, "optional" exolectins capable of attaching themselves to the ends or inside the macromolecules. Class II also recognizes two sub-classes: (1) sub-class IIa is reserved for lectins expressing an elective affinity for homotypic carbohydrate sequences such as Glu-Glu-Glu-Glu, etc., and (2) sub-class IIb for heterotypic complexes: Gl-Gal-Fru-Glu, etc. The latter are capable of contracting bonds with internal carbohydrates; they are called "Endolectins". The diagram suggested by Gallacher[68] can be reproduced as shown in Figure 10. The clarity of this option ensures its success. It allows the action mode of a lectin on a polysaccharide to be predicted and, to a certain extent, evokes the international code of enzyme nomenclature.

3.4.3.6. Toxicity of lectins

Another uncommon property of lectins is related to their possible toxicity. That of peas, for example, as commonplace as it may be, is shown to be toxic *in vitro* for certain cellular lines. An outstanding example of the danger of transposing observations made *in vitro* to natural conditions! Peas are eaten, risk-free!

In 1980, Reitman et al.[70] studied a mutant that appeared in a cellular line descending from the murin lymphoma BW. 5147. The cells resistant to the lectin are differentiated from sensitive cells by a low proportion of fucose in the heart of the glycoprotein complexes. The toxicity of abrin, an extract from the jequirity bean, is one of the oldest known since it was demonstrated in 1884 by Bruylants and Venneman[41], as previously mentioned. Such cytotoxic

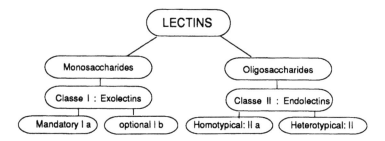

Figure 10.   Breakdown of lectins.

properties are used for selecting cellular lines. The elements are cultivated in the presence of the substance chosen for its toxicity. It deletes the majority of sensitive, so-called "normal" elements and enables the resistant ones to proliferate and survive. Everything takes place as in the selection of bacteria by antibiotics (J. Brisou,[71] 1952).

The mutants are then sampled, separated from the mass and again cultivated and cloned. If need be, it is possible to couple them with other lines that lead to the constitution of stocks of "lectin-resistant" cells. These remarkable cells are characterized by modifications dealing with surface glycoproteins. By using the selective power of the ricin lectin (*Ricinus communis*) on BHK 21 cells from the kidney of young hamsters, Hugues[72] isolated an interesting family of such resistant cells (1983). The lectin exerts its toxic power by attaching itself to beta-galactosyle terminal residues located on the oligosaccharide complexes of glycoproteins. It causes a deviation of the membrane mechanisms and provokes substantial lesions. If the initial bond can be avoided, the toxic effects are either clearly lessened or inhibited. This is an example of the importance of adherence, of its deciding role in pathogenicity.

It has been suggested that the mutants are effected by enzymatic deficiencies notably dealing with glycosyltranferases and synthetase nucleotides. It has also been put forward that the alterations effect the external chains of the complexes, including in their structures of sialyles, galactosides, and *N*-acetylglycosamines. Moreover, anomalies were pointed out on the mucins and glycosphyngolipids. Based on the above, the relatively simple idea came about, leading to the conclusion that these mutants are marked by numerous membrane alterations that give them resistance. The term "modification" is perhaps more suitable than "alteration". It is, in fact, an event that is beneficial for the cells, whereas the term alteration implies the idea of a harmful action.

It is known that saccharide structural modifications are frequent on tumoral cellular coats. Proteins and lipids are, as a whole, much less effected than polysaccharide complexes. This modification of glycolipids in malignant cells is, all in all, relatively commonplace. Research indicates that analogous events take place on membrane glycoproteins and glycosamino-glycanes.

From all these acquisitions, the result is that a veritable "membrane glycosylation" pathology could be imagined. It can be asserted that any disorder occurring on the structures as well as any dysfunction in the mechanisms of membrane biosynthesis is represented by modifications in the behavior of microorganisms in the face of receptors or even in contact, or upon approaching inert interfaces, when these variations deal with the microbial coats themselves.

Studies published by Kocourek and Hofejsi[73,74] in 1981, then by Dey and Pridham[75,76] in 1982 call for widening the definition of lectins to the point of returning to the ideas previously formulated by Bruyants and Venneman[41] in 1884. These Belgian authors compare phytoagglutinins to enzymes. Contemporary research devoted to the lectin of *Vicia faba* (the broad bean we eat) showed that it does indeed behave like an alpha-galactosidase (Kocourek et al., Shannon and Hankins, Shelinger and Schramm). It is now accepted that in certain conditions of observation, certain enzymes demonstrate hemagglutinating properties and have several bonding sites with sugars. They recognize these carbohydrate patterns and cause agglutinations. It is therefore only natural to compare them to lectins and speak in terms of "Lectins-Enzymes". Another substance of tetrameric structure was extracted from the *Vigna radiata*; it too is hemagglutinating for carrier cells where D-galactose is found. On the other hand, positive reactions were obtained with rabbit erythrocytes. The process is reversible, submitted to pH at the temperature and concentration in enzymes. This all complies with the laws of Michaelis.

Other tetrameric galactosidases were extracted from plant seeds which also recognize and agglutinate cells. As a result, the dividing line between enzymes and lectins becomes increasingly more blurry; therefore, in order not to stray, specialists are careful in the interpretations and, for the time being, only give the name "enzyme-lectin" or "lectin-enzyme" to the alpha-galactosidase of the broad bean. This enzyme is coded EC.3.2.1.22. The following can be distinguished in that of the broad bean: a simple fraction I and two fractions II1 and II2, namely three components. They are glucoproteins greedy for D-glucose/D-mannose residues. The wealth of fraction I in mannose was noted; the tetrameric configuration of the whole seems to correspond to a family of very close compounds, unions of identical monomers drawing their originality from their degree of glycosylation or from minor translations on the polypeptide chains. It can be concluded that there are three forms of this *Vicia faba* enzyme-lectin. The D-galactose behaves as a competitive inhibitor (Dey et al.,[76] 1982). This lectin therefore acts, without a doubt, as a hydrolase and a hemagglutinin. Such is its originality, but it would be surprising if it were the only one. These ideas considerably widen the framework of ectobiology.

3.4.3.7. Molecular weight of a few lectins

Abrine	132,000–135,000 for hemagglutinin
	63,000–65,000 for the toxin.
Arachis hypogea	106,000–110,000
Bandeiracea simplicifolia	114,000
Canavalia ensiformis	52,000–104,000
Dolichos biflorus	113,000
Helix pomatia	79,000
Limulus polyphemus	335,000–400,000
Phaseolus vulgaris	126,000
Ricinus communis	120,000
Solanum tuberosum	100,000
Ulex europeus	31,000–65,000 (A fraction II:weight 170,000)
Vicia faba	47,500–53,000

These weights oscillate between 50,000 and 160,000 with a few exceptions, one of which is given by *Limulus polyphemus,* the lectin of which reaches 400,000 Da. The level of carbohydrates in the molecules range from 1.3 to 11% with a more noticeable level of 21.7% for a lectin extracted from the *Ulex europeus* (fraction I).

The most common sugars are glucose, galactose, mannose, and N.Ac. Glucosamine is practically always present. Fucose, xylose, and arabinose (very rare: ricin wheat germ) are encountered less frequently. The amino acids of the protein fraction are represented by cystine and methionine, therefore thioamine acids, which deserve attention.

The affinity of the –SH groups for metals is known, therefore leaving open the possibilities of reactions, even neutralization of these complexes by the salts of bivalent cations. In addition, a substantial number of lectins have Ca^{2+} and Mg^{2+} cations which play a role in their structures, in the same way as these metals in metalloenzymes. Most often, both metals are present in a single lectin, such is the case for concanavalin, lectins from snails, lentils, Lima beans, ordinary beans, gorses, and wheat germ which also contains Zn^{2+}.

Reference should be made to the excellent work published in 1986 by Liener, Sharon, and Goldstein[63] devoted to these lectins; however, only a modest glimpse can be given here since this work is some 600 pages long!

3.4.3.8. Affinities of a few lectins

A substantial report by Albert M. Wu and Anthony Harp[77] was published in 1985 on the affinities of lectins. Reference should be made to their work for details.

A few examples will be given here: the abbreviations indicate the inhibition reactions of hemagglutinins, precipitation inhibitions, or competitive bonding tests.

Tridacna maxima
D Gal NAcl
D Gal -1-6 Gal, D Gal -1-4 Dglc, and D Gal
D Gal -1-6 Dglc, Raffinose and stachyose.

Specific lectins of sialic acid:
Agglutinin of *Limulus polyphemus* : LPA
Agglutinin of *Homarus americanus* : L Ag-1
Agglutinin of *Limax flavus* : LFA

A few lectins extracted from sponges:
Axinella : glycoprotein, specificity still undetermined.
Aaptos papillata :	N-Ac. galactosamine
Axinella polypoldes :	D-galactanes
Geodia cydonium :	Lactose and D-galactanes
Halochondria panacea :	D-galacturonic acid

It should also be noted that the *Halochondria panacea* lectin activates the growth of *Pseudomonas insolita* (Muller et al.[78] 1981).

3.4.3.9.   Examples of lectin affinities for bacteria

Lectins	Sensitive germs
Concanavalin	*AS. aureus, S. mutans, S. typhimurium, Treponema pallidum*
Wheat germ	Varied micrococci, *Neisseria gonorroheae,*
	N. meningitidis, E. coli, Treponema pallidum
Soy bean	*Streptococcus C, N. gonorroheae, B. anthracis, Tr. pallidum*
Limulin	*S. aureus, N. meningitidis, E. coli,*
Helix pomatia	*S. aureus, Streptococcus C, B. mycoides, E. coli,*
	S. typhimurium

These examples drawn from the literature provide a more accurate idea of this variability in the affinity of lectins for bacteria.

3.5.   Glycocalyx

These polysaccharide apparatuses fall quite naturally into this vast class of structural adhesins. They are fibers implanted on a large number of epitheliums and already well described in the classic treaties. They were perfectly studied and photographed on animal mucosas (intestines of cats, rats, etc.)

Ian W. Sutherland,[79] Roth's colleague, evidenced these elements in bacteria observed directly in their natural environment (water, sand, sediments, etc.). These facts have been widely confirmed by all the current studies. The authors

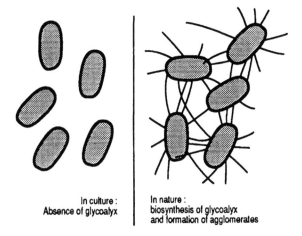

In culture :
Absence of glycoalyx

In nature :
biosynthesis of glycoalyx
and formation of agglomerates

Figure 11. Bacteria in culture and in nature. Cultures are in a very rich medium; in a minimal, very poor medium, the results may be very different.

agree, in principle, on the fact that glycocalyx are essential for the proliferation of germs in natural mediums, even on inert surfaces. In living beings, they take on a completely different significance. Unions are indeed seen between tissular glycocalyx and microbial glycocalyx. These modes of unions constitute real "bridges" or chains. The polysaccharide fibers are directed by the cells themselves which guarantee their biosynthesis, as for the other adhesins.

Onderdonk showed that *Bacterioides fragilis* adhere strongly to the peritoneum bordering the abdominal cavity of the rat, but that mutants, incapable of synthesizing glycocalyx, no longer attach themselves. Here again, is a testimony of the random character (related to genetics) of processes of adherence and behavior of microbes belonging to the same species, confronted with interfaces. On the other hand, it has been noted that glycocalyx of cells infected by a virus are sometimes modified to such an extent that their receptivity to the bacterial fibers is noticeably increased. This interesting observation sheds new light on the relations between associated bacterial and viral diseases.

Another function of glycocalyx is to preserve or concentrate the enzymes of the bacteria to guide them toward host cells. (See Figures 11–12.) In such conditions, these polysaccharide fibers behave like genuine food reservoirs available at any time for the attached microbes.

Lastly, it should be recalled that glycocalyx have the property of attaching ions and stocking them. They act the same way for all the mineral or organic molecules essential for life. The protective role of these adhesins should also be emphasized as they have the property of picking up and neutralizing certain, possibly undesirable, ions that are present either in the form of salts or incor-

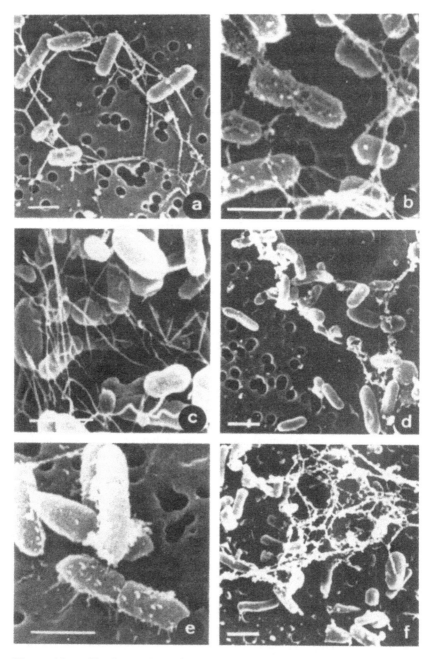

Figure 12.　*Citrobacter freundii* and *Proteus mirabilis.* Composition of the media was extremely limited. (From Richelle-Maurer, E. and Moreau, Z., *Rapp. Comm. Int. Mediterr.* 1986. With permission.)

porated in complexes filling the immediate environment. Everything is linked to the nature of such salts — their solubility in water or lipids, as well as their power of penetration, which is also subjected to laws. These polysaccharides, organized in genuine and sometimes very dense networks, have led to comparisons with spider webs or spools of thread. These remarkable elements ensure the protection of microbial cells against biological aggressions such as antibodies, antibiotics, and phagocytose. Their mission of attachment is supplemented by a mission of protection.

All these structures consolidate the cohesion of microbial colonies set up on inert or living interfaces. Many species live in perfect harmony, in synergy, or even in symbiosis as confirmed by Wolf after studying the microbial colonization of sea sediments. However, everything is not clear in this complex field. It was asserted (Costerton)[7,8] that germs do not synthesize glycocalyx *in vitro*. Finalists consider this biosynthesis as useless. Without taking sides in the debate, the studies made by E. Richelle-Maurer and Z. Moureau[80] at the Royal Institute of Natural Sciences in Brussels in 1986 should be noted. These authors observed the presence of numerous fibers compared to glycocalyx in a variety of bacterial cultures. The composition of the culture media adopted for the experiment was extremely limited; they were qualified as "Minimals", placing the germs at the limit of growth possibilities. Images obtained using electron microscopy illustrate these organizations forming large networks. However, it may be possible that such culture conditions, bordering on scarcity, encourage the bacteria to synthesize glycocalyx as in nature, which they do not do in an environment where they exist in abundance.

This important chapter devoted to adhesins will be concluded in this manner, with a glimpse at their extreme complexity. These adhesins, or anchorage organs, are sometimes of glucoprotein nature, sometimes purely polysaccharide, and sometimes associated with lipids. Some are solid, participating in the cellular structure in its coat; others are in mucoid forms, viscous liquids, or soluble substances. In this field, all the elements are dependent. On the other hand, the biosynthesis of adhesins is governed by genetics; their birth is reliant on the environment, on organisms themselves, on their physiology, and on the functioning of membrane complexes.

Such are the foundations of microbioecology which still hold many unknowns and lead to cautiousness in the interpretation of events observed in the book of nature, so difficult to read! Essential to remember is that it is always a matter of encounters and recognitions of molecules and the expression of sometimes completely unexpected stereochemical affinities. A few diagrams and reproductions of illustrations will make it easier to understand part of this complexity.

3.6. Fibrillae

Fibrillae, surface organella for a long time confused with fimbriae, are in fact very different. These very small-sized filaments were observed in *E. coli* K88 and K89. Fibrillae are inserted deeper, like sexual pili, on basal corpuscles next to the inner side of cytoplasmic membranes. On the surface of cells, they sometimes form compact, amorphic masses, or are organized in helices.

There is no doubt that fibrillae participate in the mechanisms of adherence.

3.7. Other Possible Adhesins

All the equipment described on the cellular surfaces, notably in bacteria, such as locomotion flagella and other accessories can, in certain circumstances, contribute to the mechanisms of adherence. The purely fortuitous entanglement of all these filaments can only enhance the formation of aggregates and agglutinants, intricateness owing nothing to the stereochemical processes. Walsby et al.[81] (1992) drew attention to a few cyanobacteria living in freshwater media. They were endowed with small vesicles that influence their density and play a certain role in their behavior, even their attachment, to other particles during their travels. The appearance of these microvesicles is subject to light intensity and regulates vertical migrations in the water. Other bacteria conduct their trips and, at the same time, their opportunities for encounters with possible receptors thanks to similar means. A.E. Clarke et al.[313] (1933) and Ramakrishnan et al.[82] (1992) studied the "porine" of *H. influenzae* type b. This envelope is composed of 34 amino acids. It takes up a lot of room on the bacteria surface. It is considered highly antigenic, thanks to the construction of models which enabled the calculation of the hydrophobicity, amphicity, and organization of amino acid sequences. The role of this envelope, and its role on the bacteria surface have been partially specified.

3.8. Spores

Numerous aerobic (*Bacillus*) or anaerobic (*Clostridiaceae*) bacteria sporulate and release their forms of resistance in nature. During their intracellular formation, there is substantial production of calcium dipicolinate. Once dispersed in an ordinary environment, these small spheres are submitted to the laws of all the particles; they are therefore adsorbed by soils and sediments and are deposited on all the interfaces as they wait to find conditions enabling germination and a return to an active life. Such is the case for the spores of *Clostridium tetani*, *Clostridium perfringens*, *Bacillus anthracis*, etc. On the surface of a few of them, filiform, tubular appendices with a variety of aspects, can be observed. These organella do not fail to evoke adhesins. Debate continues concerning the function of these coating accessories which only appear in the strictest conditions of anaerobiosis, and only in some species (*B. anthracis,*

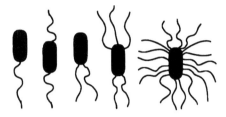

Figure 13. Various types of cilia in bacteria. These organs, mainly designed for locomotion, are nonetheless carriers of antigenic structures (H) and of lectins. The fact is proven in certain strains of *E. coli* and of *Serratia marcescens.* These organs must therefore play a role in the formation of agglomerates and perhaps in adherence. However, this remains a topic of debate, notably for cholerigenic vibrions.

for example), the surface of which is covered with filaments. It is tempting to consider them as attachment, or even adherence, apparatuses, but this is only a hypothesis. (See Figures 13–14.)

4. A FEW GENETIC REMINDERS

The concepts of genetics, some of which are already old, will only be briefly recalled, insofar as they directly interest what has just been presented. Brinton's[83] work (1965) confirmed what was known of the genetics of fimbriae and shed additional light on this knowledge. Three cistrons appear to direct the biosynthesis of type I fimbriae. The responsible genes were transferred from the chromosome on vector plasmids (Hull et al.,[84] 1982, Purcell et al.,[85] 1981). Genetic engineering uses a carrier plasmid of the gene coding a type I fimbriae and introduces it into an *E. coli* strain normally deprived of fimbriae. The carrier bacteria of the plasmid agglutinates the guinea pig erythrocytes and responds to the solicitation of the type I antifimbriae antibodies. The translated protein contains 23 residues of peptide signals, in addition to fimbrial subunits themselves including 159 amino acids. According to Brinton,[83] the bacteria endowed with fimbrial genes can spontaneously change phases and no longer function. Apparently, this is a matter of simple variation that cannot be confused with mutations. The frequency of these events is estimated at approximately 1/1000 bacteria per generation. Eisenstein et al.[86,87] reported in 1981 and 1982 that the appearance of fimbriae in salmonella observes the same laws as those governing the genetics of flagella. The variations are governed by structural genes and regulator genes, as is classically the case. The genetic support is known by the abbreviation "Fim A." (Fim A gene).

Any culture in the process of growing in which carrier bacteria of Fim A gene exist is, in fact, a mixture of phenotypes covered with fimbriae and cells deprived of these organella. The proportion of these phenotypes is related to

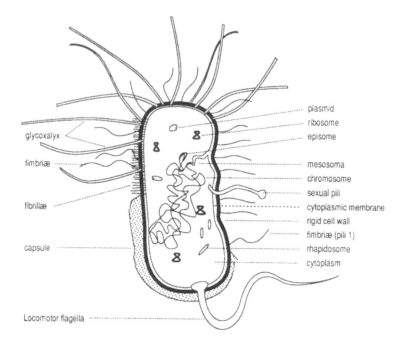

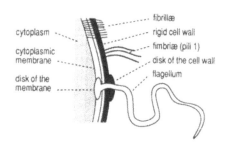

Figure 14. Top — Main components of bacteria. Bottom — Attachment of the pili, fibrillae, and flagella on a bacterium.

the nature of the microenvironment, and to different factors, some of which are favorable, while others are hostile to the expression of this Fim A gene.

Forty-eight hour cultures in stable liquid media, sheltered from any agitation, are conditions favorable for the expression of the gene; however, nutritive agars, liquid media enriched with glucose levels of 1%, penalize the biosynthesis of the fimbriae. This influence of the medium conditions on the bacterial variability raises the question of the predominance of a given phenotype over another during infections.

For now, we can state that *E. coli* isolated from urine are deprived of mannose-sensitive lectins, whereas the same strains cultivated in the laboratory produce these coating organella *in vitro*. During the same time period, a rise in the antifimbriae antibodies can be noted in the serum of the patients (Ofek et al.,[88] 1981, Haber et al.,[89] 1982, Schwartz et al.,[90] 1982). It is accepted that at the beginning of the infection, the responsible germs are rich in fimbriae and that later, under the action of immunitary defenses, they lose their adhesins. Such observations confirm the value of specific antigen treatment by autovaccinations, to which we have always given preference (E. Magrou and J. Brisou, 1947).[91] Other tests, performed on newborn rats, reveal that if animals are infected with a mixture of carriers and non-carriers of fimbriae, 24 hours later only bacteria with fimbriae are found in the buccal cavity; however, the phenotype, deprived of these accessories, is present in the blood. Similar observations were recorded with *K. pneumoniae* experimentally infecting mice (Maayan et al.,[92] 1985). All the facts mentioned here confirm that a bacterial strain can, depending on the outside influences and circumstances of all kinds, go rapidly from one phase to another. It is easy to grasp the importance in the evolution of diseases.

We know that the coding of sexual pili is essentially located in the "R" plasmids which preside over the fertility of the F^+ cells. This reminder emphasizes that genetics, as specified by S. Normark et al.[93] (1986) constitute an excellent tool for studying the mechanisms of adherence. *E. coli* strains are endowed with multiple gene-coding adhesins of varied specificities and expression. Many studies have been devoted to the recognition, isolation, and count of DNA sequences, cistrons responsible for the biosynthesis of adhesins in general. Mutations have succeeded on genes. The specific function of each one has been specified, notably as far as the synthesis of pili are concerned and their function in the adherence process. Orkov et al.[94] (1975) recognized very close structures on *E. coli* K.99 strains isolated from calves and lambs suffering from diarrhea. The K.99 antigen is mannose-resistant with everything that entails with respect to erythrocytes in various animals. They are no longer agglutinated by such strains. It is assumed that the specific receptors of the K.99 antigen can correspond to a terminal *N*-acetyl-galactosamine and residues of sialic acid, glycophorine, and other glucoconjugations (Lindahl et al., 1984).[95] Those responsible for the coding of the antigen are located on plasmids. Conversely, the pili of *E. coli* 987 which are numerous, are submitted to chromosomal genes. The same is true for the enterotoxic strains, isolated in human carriers of the CFA.I and CFA.II antigens.

Another aspect of the role of genetics in all these phenomena is provided by Steidler et al.[96] (1993) when they show that the fusion between a Pap A gene and fragments coding an immunoglobulin G in a strain of *S. aureus* is produced in such a way that certain fragments (Spa) are found inserted in a segment of the codon 7 or 68 governing the nature of the Pap A. The peptides of this zone are located at the extremity of the pili. The alteration of the bacterial strain operon makes it possible to specify the area of activity of a bond between an immunoglobulin G and the surface of the microbial cells.

The uropathogenic, mannose-resistant *E. coli* are equipped with indifferent fimbriae named "Fimbriae P", or "Bonding Gal-gal-pili", "digalactosidic bond pili", "galactosidic bond pili", or lastly "Pap pili" (Pili-associated pyelonephritis). There are, of course, several serotypes of these Pap-pili. Genetic engineering specialists implement all their skill to create new strains, with a determined specific property, uncommon adhesins, a surprising virulence, or a toxigenesis. All kinds of combinations are possible.

Staffan Normark et al.[97] (1983) were particularly interested in the uropathogenic *E. coli*. Based on daily laboratory experience, we know there are mannose-sensitive (MS) strains and mannose-resistant (MR) strains. The genic group coding the type I pili is located on the bacterial chromosome and not on a plasmid. The majority of strains have these genes, irrespective of the origin. On the other hand, the genes of the strains responsible for epizootic diarrhea in piglets are located on plasmids. These strains are mannose-resistant. The K.88 antigen of *E. coli* is directly responsible for virulence. It contributes to the attachment of the enterotoxigenic bacteria. The intestinal mucosa of the young animal, indifferent to the mannose, also makes it insensitive to agglutition, which leads to the conclusion that its affinity is addressed to different receptors. (See Figure 15.)

Non-digalactosidic MR adhesins are currently designated by the letter X, as X-adhesins. They are found in the *E. coli* strains isolated from non-intestinal infections. *N*-aceytl-glucosamine and sialylgalactosides seem to serve as specific receptors for some of these strains (Parkkinen et al.,[98] 1983). It was possible to clone such adhesins descending from *E. coli* isolated from pyelonephritis and upper urinary infections. The 52 Pap-pili associated with these urinary infections are, naturally, genetically coded. A certain number of variants have been described confirming the extreme complexity and diversity of the surface antigens of microorganisms in general. Two major antigens descending from the pili are designated by the abbreviations IC and F.13. The F.13 antigen appears to have a close neighbor abbreviated as F.12. The MS and MR adhesins have their very specific, characterized genetic site. They are related to diagalactoside. Various studies indicate approxiamtely 8.5 kilobases of cloned DNA are required for the construction of a single Pap-pili, related to di-galactoside. This coding element is designated here again by an abbreviation: Pap-DNA. In fact, it carries eight genes. The Pap-A codes a subunit of pili (the F.13 antigen mentioned above). Attempts have been made to locate this Pap-A and to try to understand the organization of the group of genes responsible for the di-galactoside coding specific to the adhesin.

We owe a large part of this outstanding knowledge to the meticulous research by Van Die mentioned by Normark et al.[99] (1986). All these facts fully confirm and complement previously known ideas, notably those that were provided by Evans[100] in 1978; he evidenced the plasmidic coding of the CFA/I

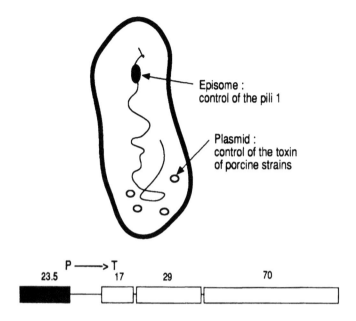

Episome :
control of the pili 1

Plasmid :
control of the toxin
of porcine strains

P ——> T
23.5   17   29   70

Figure 15. **Genetic sequence of the biosynthesis of *E. coli* K88 adhesins.**

factor of enterotoxigenic *E. coli*. It should be specified that two sites intervene to guide the biosynthesis of this antigen. The thermosensitive toxin is coded at a currently identified level, responsible for the organization of a minimum of six polypeptides including two that participate in a fimbrial subunit. The same is true for numerous fimbriae and classic bacterial antigenic structures.

In certain mutants, the Pap gene beam no longer provides the progress of the bonding phenomena, but it preserves the power to code the biosynthesis of the Pap-pili. By carrying out a controlled series of cloning, artificial mutations, and recombinations, geneticists showed the substantial complexity of these mechanisms, the position on the genes of the sites responsible for the construction of sub-units of pilin, specific adhesins, receptors, etc.

All these genetically controlled elements are essential to the final completion of irreversible bonds. A particular *E. coli* strain, I.96, served as a research tool as far as the previously mentioned digalactosidic bonding adhesin is concerned. The question raised was to know whether or not the genetic responsible for adherence could be eliminated without affecting the production of the Pap-pili. The conclusion of this research was that although certain genes appear to be indispensable for the expression of the adhesin, they are not essential for the formation of the Pap-pili. Without going into the technical details, it is worthwhile to recall a few essential ideas. On a single bacterium, it is possible to locate 8–10 different regions likely to code adhesins and pili. For example, a region located between the Pap-A and Pap-F loci can also be a carrier of three cistrons coding Pap-H, Pap-C and Pap-D (Baga et al.[101]). The Pap-H gene alone codes a protein of 20 kDa! The expression of all these genes is indispensable for the construction of the Pap-pili which governs the specific digalactosidic bonding activity.

The proteins coded by all these genes have been the subject of some very in-depth analyses, notably due to autoradiographies. These proteins share a certain number of amino residues with other pilins. Among the latter, two cysteins and a penultiem, tyrosine residue, are noteworthy.

Purified Pap-pili enabled the preparation of specific antiserums (Lindberg et al., 1986, mentioned by Normark).[99] There are 98% of Pap-A pilin subunits in the mixtures. It was thus concluded that the Pap-A gene had dominant efficacy. It is attributed with a length of 700 bases.

An MR type non-digalactosidic adhesin was isolated from another *E. coli* strain 08:K15:H31.UTT. The clone expresses the pili agglutinating the bovine erythrocytes and at a much lower rate, the human red cells. Subclones were isolated coding the one responsible for agglutination but not the one for the formation of the pili. The location of a certain number of genes governing the synthesis of specific, responsible functions is now guaranteed. This is the case, for example, for the enterotoxigenic *E. coli* adhesins K.88 and K.99. These bacteria are especially of interest to veterinary medicine, but there are so-called CFA strains of human origin. This knowledge is a result of the studies summarized by Gaastra and De Graaf[102] in 1982. We know that the code of K.88 adhesins is located on plasmids which enable genetic conjugations. These plasmids are capable of transporting the faculty of the bacteria for using raffinose as the only source of carbon (so-called Raf+ strains). Gene beams were cloned by Mooi[103] in 1979, and by Shipley et al.[104] in 1981. They contain six elements. The genetic map of cloned DNA of Pap K.99 and K.88 ab has been drafted. The CFA/I adhesin is coded by plasmids which do not enable conjugations. These same plasmids often take care of directing the biosynthesis of thermoresistant toxins. Another very specific plasmid, NTP.113, expresses the toxin ST.CFA/I. Two regions were identified, and both were recognized as indispensable for the appearance of the CFA/I character. They were cloned. The ST gene is very closely related to the sector coding the subunit CFA/I. Such knowledge related to the adhesins of pili and to the biogenesis of Pap-pili of digalactosidic bonding suggests that the proteins of Pap-A, H, E, and F correspond to apparently normal peptides in the *E. coli*. They follow the usual path of secretion through the cytoplasmic membrane. However, the exact location of this pilin pool preceding the polymerization that will lead to the final organization is not yet known.

Some authors (including Bayer, mentioned by Normark) advance the hypothesis that sites outside and inside the membrane may merge. A flux appears to be established between the reversible phospholipids and irreversible lipopolysaccharides, using the bonds suggested by Bayer. Nothing is opposed to the idea that these still unorganized sites correspond, in the end, to the points of formation and construction of the pili. Specialists consider that the hydrophobic C-terminals of most of the pilins are indispensable for polymerizations.

In this way, the Pap-D protein can form a complex with different pilin Paps and release them from the cytoplasmic membrane, thus protecting them from a hydrolasic action that would be catastrophic.

Such cooperation between these pili construction mechanisms is perfectly remarkable. Everything is coded and coordinated at these molecular levels. The pili are born together, descending from a joint pool of materials, all benefitting from temporary protective unions and escaping undesirable aggressions of hydrolases during the course of their ontogenesis. They express and state their originality once the obstacles and various stumbling blocks have been avoided on the membrane. Some examples: in a Pap-E mutant, the Pap-F can be polymerized in a Pap-C/Pap-A complex. The Pap-A pilin subunit gives the structure to the pili, but we still do not know the exact position as far as the minor subunits are concerned. Some authors suggest the possibility of a co-polymerization phenomenon, occuring during the construction of the pili on both sides of the membrane.

Henry and Pratt,[105] (1969) studying the biogenesis of the filaments of the bacteriophages fl, fd, and M13, had already noted the presence of a minor coating protein called "protein A", which was essential for ensuring the attachment of the virus on the host bacteria. Everyone knows the role of certain lectins in this process. It is also known that amino acids are themselves indispensable as well. Such is the case of tryptophane for the attachment of coli-phages. Other complexes have been identified at the surface of virions, notably glycoproteins. It is currently accepted that the bonding of two types of organisms is a result of coordination, joint actions between pilins, and viral filaments.

The Pap-G protein is indispensable for the union of intact cells and purified pili (Norgren et al.,[106] 1984, Van Die et al.[107] 1984). It is a precursor located on both sides of the membrane. It is considered as a genuine adhesin.

From this chapter, we are intentionally excluding the role of sexual pili called "F" or fertility pili, also coded, but the very specific functions of which move away from adherence as it has been defined and understood. It is obvious that the exchange of genetic material requires a solid union between both partners in prokaryotes, but this is an event that has nothing to do with what is being analyzed here.

The MR *E. coli* adhesins are also coded by gene beams that are not found in the commonplace Escherichia belonging to the intestinal microbism. The responsible DNA is located on the chromosome. In some uropathogenic strains, the coding of hemolysine and pili-adhesins is linked to a series of apparently fortuitous events. The intervention of character transposition mechanisms on the genes responsible for hemolysines, called "Hly" (hemolysines) "MRHA" (mannose-resistant-hemagglutinants). According to Low et al.[108] (1984), if these two factors Hly and MRHA act in synergy or in complementarity, they must cause a strong selective pressure to stop both properties. As the Hly and MRHA determinants are coded by plasmids and located on the same element, then integrated in the chromosome, perhaps they acquire the power to express their original bonding character.

5. CONCLUSION

The reader, confronted with a few concrete examples which will be examined in the second part, should understand that these fimbriae-lectins or structural adhesins are necessarily submitted to the laws of genetics, due to their very nature. Bacteria, being prokaryotes, have a coding that is sometimes plasmidic and sometimes chromosomal. What we know about transfers in the bacterial world leaves room for numerous means of expression of the properties that interest us and provide each strain with its own identity.

In eukaryotes, the construction of adhesins abides by the same laws of chromosomic genetics, here complicated by mitoses. With bacteria, we have to count on the intervention of genetically dependent factors, the role of which is capital for the construction of a single pili. We are not yet perfectly informed of the way in which the proteins act in relation to each other. Some uncertainties burden the comprehension of regulation mechanisms which end in the completion of final structures. The extreme complexity of the mechanisms justifies the reserve that has been expressed for years on the topic of "Bacterial Species".

2   CHEMICAL ELEMENTS OF ADHERENCE

1. THE BACTERIAL WALL

With the exception of the *Mycoplasma*, all bacteria are protected by a thick wall. These *Mycoplasma* are no less deprived of adhesins, which were studied in depth by Razin et al.[314,315] in 1981 and by Bredt et al.[316] the same year. These exoskeletons left their marks in extremely old sediments. Commonplace in appearance, this envelope conditions the morphology of the cells. It is only understood through biochemistry, which explains why it has been considered at the beginning of this chapter. This wall offers points of support to all the organs we just inventoried as adhesins and also resists very high pressure due to its rigidity. As early as 1884, Adrien Certes[109,110] had already made a note of this fact. Oppenheimer et al.[111] (1952) emphasized the selective role of these pressures. They studied barophilic and cryophilic species isolated in marine sediments extracted from depths ranging from 10,000 to 10,450 meters, in the abyssal cavity, the Mariana Islands. Common species can be encountered at 4000 or 5000 meters in depth or even more in oceans, as we demonstrated in 1955 in a work devoted to marine microbiology (Part 2, J. Brisou et al.[321]).

Despite its apparent rigidity, this wall preserves a certain amount of flexibility and substantial elasticity. The structures are rich in antigenic patterns in varied receptors, some of which let the bacteriophages pass. This barrier also plays a considerable role in the response given by the cells to the antagonist attacks such as antibiotics and other antibacterials. The sensitivity of the germs to one coloration method or another is intimately linked to the structures that make it possible to differentiate the Gram-positive bacteria from the Gram-negative cells. These well-established ideas will not be developed here.

There are numerous contact points at various locations between the bacterial wall (a genuine cell wall) and the cytoplasmic membrane. These contacts facilitate permanent exchanges between the two envelopes. Peptides, polysaccharides, lipids, and other complexes constitute the essential part of the chemical structures that will be examined, endowing each genus, each species, and each variety with its originality. The most significant complexes are represented by peptidoglycans where murein and its derivatives predominate. Even a limited biochemical analysis of these parietal components and adhesins guides the choice of specific enzymes used during release tests. These compounds are indeed "substrates" that are worth knowing.

2. POLYSACCHARIDES

Since 1884, following Emil Fischer's[112] work, the structure of glucose has been established as a point of departure for studies devoted to the chemistry of carbohydrates and polysaccharides in general, which are broken down into two groups: oligosaccharides and polysaccharides, depending on the length of the chains. Their position, either inside or outside the cell, also makes it possible to distinguish *endo* and *exo* polysaccharides. All of them are sensitive to hydrolases known by the name of glucosidases which are highly numerous and useful to know about for the releases. They will be analyzed in the chapter devoted to techniques.

These polysaccharides are obviously constituted of chains of monosaccharides with polyalcohols sometimes including an aldehyde function and sometimes a ceton function. They represent the most substantial components of the biosphere as demonstrated by cellulose or starch. Depending on their structure, whether *endo* or *exo* cellular, homopolysaccharides and heteropolysaccharides should be distinguished. Moreover, varied residues, cetons, and acyl groups complicate these molecules by endowing them with chemical and stereochemical specificity: alginates are homopolymers of D-mannuronic acid and cellulose is a homopolymer of glucose units. Exopolysaccharides of *Klebsiella* have been the subject of numerous studies: Erbing et al.[113] (1976), Dutton[114] (1973–74), and Thurow et al.[115] (1975).

Depending on the strains studied and the authors, fucose, glucose, pyruvate, or glucuronic acid-based complexes were identified, as well as mannose and galactose, united with glucose and pyruvate. Glucose, glucuronic acid, pyruvate, and galactose count among the most usual components of Gram-negative bacteria. In the single group of *Klebsiella*, 81 combinations were enumerated. (See Figures 16–17.)

2.1. Biosynthesis of Polysaccharides

The biosynthesis and molecular modifications of these complexes take place due to the intervention of nucleotides, phosphate esters, and cyclitols, constituting an important class of intermediates. Cyclitols, for example, transform hexoses in uronic acids. The phosphoric esters participate in the structures of nucleotides and in energy sources. In 1906, Harden and Young[116] discovered

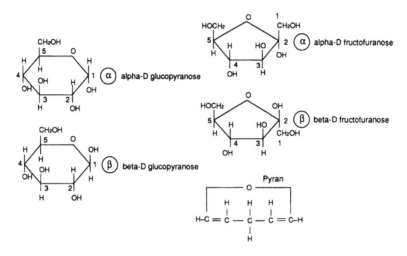

Figure 16. Basic structures.

the phosphorized esters of glucose. These studies were the departure point for all the research devoted to nucleotides and nucleosides. Following is a list of them and their corresponding abbreviations:

UDP	uridine diphosphate
UTP	uridine triphosphate
UDP Glc	uridine diphosphoglucose
UDP Gal	uridine diphosphogalactose, etc.

The biosynthesis of these molecules is obviously governed genetically, as well as by enzymes belonging for the most part to "glucosyl transferases". In general, microorganisms do not participate in the synthesis of glucose; instead, they are reserved to vegetals and hepatic and renal cells as well as a few cells of the small intestine in animals. The biosynthesis of polysaccharides is, however, widespread in the bacterial and microbial world in general. Prepared sugars, conditioned by a transferase, are polymerized. Other transferases free glucose, for example, by attacking saccharose to then polymerize fructose and produce fructosans. Extracts of *Neisseria perflava* catalyze this reaction as follows:

$$\text{Saccharose} \rightarrow \text{Fructose} + \text{Glucose}$$

due to a beta-D-fructofuranosidase E.C.3.2.1.26. The fructose is taken by other enzymes of group 2.7 belonging to the class of transferases. Certain biosyntheses take place without the participation of phosphorus:

$$\text{Glucosyl-I-glucoside} \rightarrow \text{Polysaccharide}$$

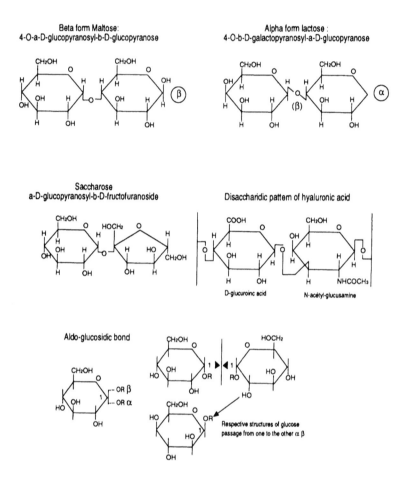

Figure 17. Complex structures.

The energy needed for the release of glucose requires the intervention of ATP, but the continuation of the operations is ensured by an exergonic source of substitution.

From the already-old work by Beijerinck[117] (1910), it should be retained that certain bacteria synthesize polysaccharides using isolated glucose, but they are not capable of acting in a simple mixture of glucose and fructose. Bacterial metabolisms offer other examples of these singularities. *Leuconostoc mesentericus*, for example, synthesizes polysaccharides very easily in the absence of phosphates and directly ensures this same reaction:

$$\text{Saccharose} \rightarrow \text{Fructose} + \text{Glucose}$$

A dextransucrase governs the synthesis of molecules of 75,000 Da in the presence of other sugars such as maltose, isomaltose, and glucose. Numerous enterobacteria synthesize specific sugars by following slightly similar pathways, with the help of transferases. Colitose, abequose, tyvelose, and paratose are born in this way.

It is reasonable to apply the same rules to the biosynthesis of glycocalyx, the polysaccharides of which constitute the essential part, as well as to capsules. These biosyntheses are obviously subjected to numerous factors, and are in close relation to the vital elements of the microenvironment. Such is the case for the content in oxygen, in growth factors, or in trace elements, etc. It has been established that in the presence of saccharose, the production of alginate by *Arthrobacter vinelandii* is hindered by an excess of oxygen or overly energetic agitation. In such conditions, bacteria consume too much sugar or produce more CO_2 to the detriment of a more active biosynthesis of the polysaccharide (T.R. Jarman et al.,[118] 1978). The presence of certain trace elements M_0O_4, PO_4, and Fe_2^+, activate the synthesis of this alginate. It would nevertheless be an exaggeration to generalize the value of these experiments performed *in vitro*.

The precursors for the biosynthesis of polysaccharides are nucleosides and cofactors of polyisoprenosyl phosphates which intervene at the same time in the synthesis of parietal elements. The intervention of competitive actions likely to act on bacteria were suggested, either in favor of the construction of walls or to the benefit of a direct production of polysaccharides. This complex process continues to divide specialists. Before undertaking a rapid assessment of the main polysaccharides likely to interest us, here is a classification based on basic monosaccharides. (See Figure 18.)

2.2. Main Bacterial Polysaccharides

2.2.1. Levans

Levans are polyfructosans with fructose for a basic element. They are synthesized by numerous bacteria in which the bonds follow the beta model $(2 \rightarrow 6)$. It is accepted that bacterial levans are made up of long chains resulting from the transfer of beta fructofuranosyl from a donor on a receptor with, for a departure point, saccharose, the fructose of which has been released. The hypothesis is checked by the presence of glucose among the components. An average of 9–10 beta $(2 \rightarrow 6)$ fructosans linked with numerous ramifications constitute all the molecules of levans. Among the concerned bacteria, it should be noted above all that *Pseudomonas*, *Alcaligenes viscosus*, *Aerobacter levanicum*, *Streptococcus salivarius* and numerous *Bacillus* clearly dominate the levan producers.

2.2.2. Polymannans

These macromolecules are built of mannose units, as their name indicates. They are encountered not only in bacteria, but also in lower fungi. The bonds

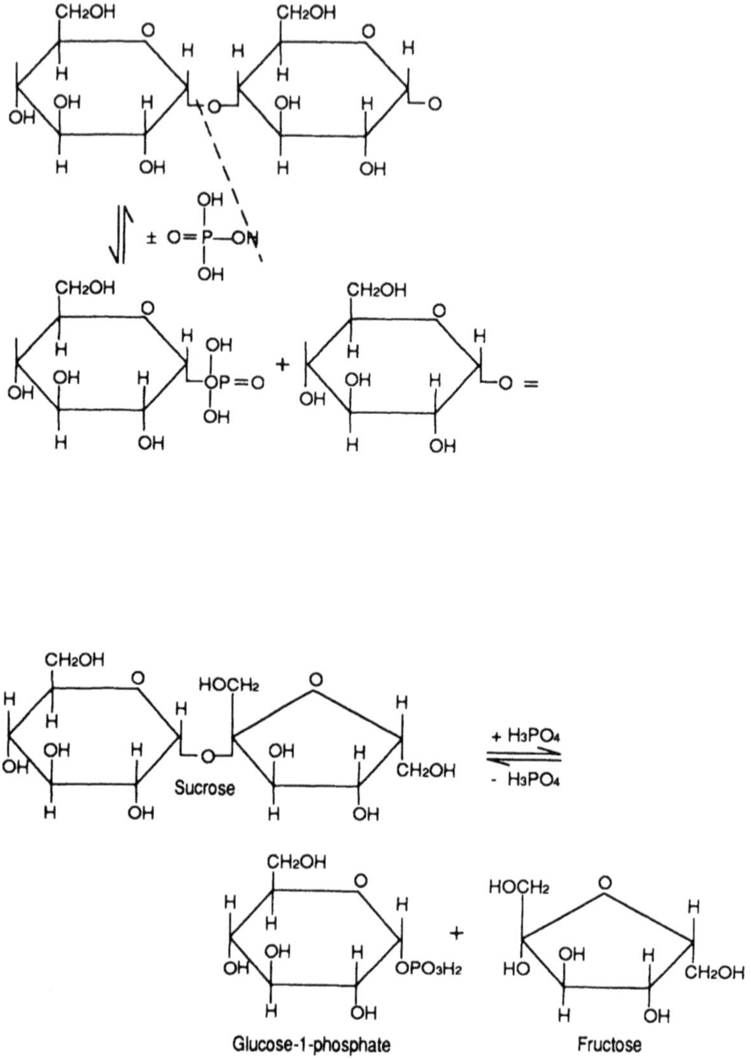

Sucrose

Glucose-1-phosphate   Fructose

Figure 18.   **Top: Mode of action of phosphorylase E.C.2.4.1.1. on a non-reducer end, during the activation of a polysaccharide. Bottom: Reaction of sucrose phosphorylase E.C.2.4.1.7. or sucrose glycosyl transferase sucrose-orthophosphatase, alpha-D-glucosyltransferase.**

are of the $(1 \rightarrow 3)$ type. Hydrolysis delivers 2-O-alpha-D-mannopyranosyl-D-mannose, a 2,3,4,6 tetra-methyl D-mannose, and a 3,4,6 tri-O-methyl-D-mannose. Analyses make it possible to detect small quantities of 2,4,6 tri-O-methyl-D mannose and 3-O-methyl-D mannose. These results suggest a main $(1 \rightarrow 2)$ type structure and bonds $(1 \rightarrow 6)$ and D-mannopyranosyl forming

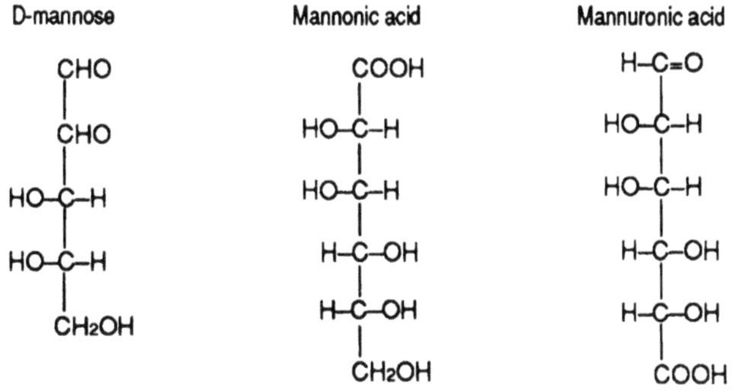

Figure 19. Structure of D-mannose and derivatives. This non-reducer sugar enters into the structure of mucoid substances. It is one of the main polysaccharides of the tuberculoprotein. The bacteria are producers of mannosanes, and we pointed out the importance of D-mannose during certain inhibitions of adherence. There are mannose-sensitive and mannose-resistant bacteria (E. coli, for example).

units, located in ramifications on the main chain by bridges (1 > 2). (See Figures 19–20.)

2.2.3. Dextrans

This term englobes a class of polyglucans essentially composed of Alpha-1,6-D glucopyranosyl, as far as the main chain is concerned. These dextrans serve as reserve materials for the microorganisms. Numerous bacteria are producers of them, notably lactic bacteria. The synthetic enzyme, a glycosyl transferase, is a component in Streptococcus mutans, whereas it is induced in Leuconostoc melanoïdes.

Certain dextrans are soluble, whereas others remain indifferent. The first are attributed with the viscosity of some biological liquids, promoting bacterial agglomerates. Two variants can be simultaneously synthesized by a single bacteria. Leuconostoc mesenteroides, a very active producer of this polymer, is widely used in industry and the marketing of these dextrans. The yield of biosynthesis is accelerated by the presence of saccharose. The complexes marketed are the alpha-1.6 type with lateral chains. Moreover, they are valuable antigens (Allen and de Kabat[119]).

When other sugars are present in the medium, which is common in nature, the enzyme synthesizes derivatives, of which "leucose" is an example. The role played by the dextrans in the attachment of bacteria is considerable, notably on the solid interfaces. Examples will be given in the constitution of biofilms on teeth. Quite a few bacteria such as Acetobacter capsulatum, Streptococcus bovis, Streptococcus viridis, Streptococcus viscosum, Leuconostoc dextanicum,

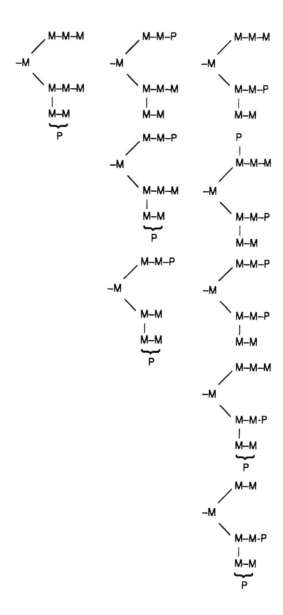

Figure 20.   **High-density structure of phosophrylated mannose detected in the murine lymphoma BW 5147.3 by Varki and Kornfeld in 1980.**

Betabacterium vermiformis, Betacoccus arabinosaceus, Streptococcus dextranicum, etc., all offer proof.

Electron microscopy makes it possible to specify the aspect of these structures. Whereas *S. mutans* synthesizes glucans, *Streptococcus sanguis* exclusively reserves the dextrans. These polysaccharides are rich in elipsoidal particles forming aggregates in the heart of which the beginnings of fibrillae are detected. These were qualified as "profibrillae" by Guggenheim and Schroeder[120] in 1967. This fibrillar nature of the polymers synthesized in the presence of saccharose was confirmed in 1971 by Johnson et al.[121]

2.2.4. Celluloses

These equally multiple polysaccharides represent by themselves over 50% of the total carbon in the biosphere. Bacterial celluloses are always extracellular, which is extremely noteworthy in the release tests. They are homopolysaccharides resulting from glucose chains. Cellobiosis, diholoside of the celluloses, is characterized by its reducing power. Acid hydrolysis releases compounds of D-glucose, whereas an arranged action leads to cellobiosis. The glucose units are united in beta (1–4). The weights oscillate between 300 and 3000 glucose units.

Lipid intermediaries are involved in the synthesis of these polymers. The cellulosic networks are, in general, tight, and the fibers are cemented by hemicellulose, pectin, and extensin. The hemicelluloses themselves recognize the following as components: Beta-1–4-D-glucan, D-galactose, D-mannose, beta-1—3-D xylan, L-arabinose. The heterogeneity of the group is thus confirmed. This cellulose is often given as an example of a polysaccharide synthesized by bacteria as well as by prokaryotes, in general, and eukaryotes, but it obviously is not an exclusivity. Other polymers present the same character.

2.2.5. Crown Gall Polysaccharide

The Crown Gall polysaccharide is a specific complex produced by *Agrobacterium tumefaciens,* the agent of Crown Gall. This polyoside is totally hydrolyzed by HC_1-N at 98°C in an 1.50 h. It releases D-glucose and D-glucobenzidazole. Beta-gluconic bonds dominate the structure of this compound, the molecular weight of which is close to 3500.

Agrobacteria, with numerous biovars, do not have the exclusivity of synthesizing these glucans. *Bradyrhizobia* are also good producers of them. The polysaccharide of *Agrobacterium* is hydrosoluble. Beta-glucan remains the main element, but succinic, pyruvic, and acetic acids are also encountered. This structure is rather close to that of the heteropolysaccharides of *Rhizobium. Agrobacterium tumefaciens* is responsible for tumors developed on certain trees; for this, it develops cellulosic fibers playing a probable role in its attachment on vegetal cells.

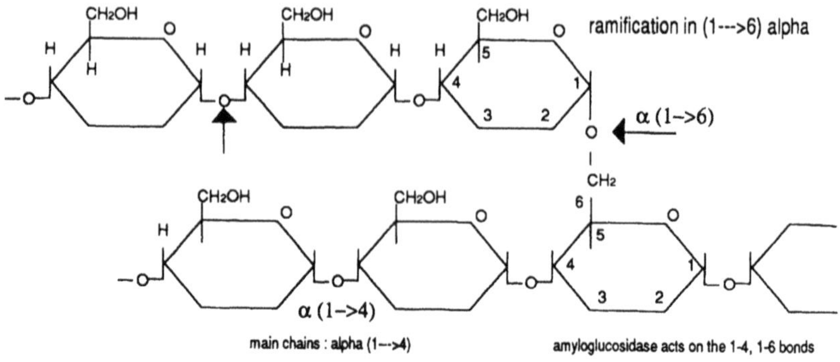

Figure 21. Structure of amylopectin.

2.2.6. Starch — Amylopectin

The most substantial starch of natural carbohydrates is essentially a veg-etal product. It exists in two forms: alpha amylase and amylopectin. The first is composed of long chains of glucose, whereas the second presents numerous ramifications. Although the starch is hydrolyzed by two enzymes, alpha amy-lase and beta amylase, amylopectin is itself also sensitive to these same two hydrolases, but with a particularity as far as its complete transformation is concerned. Amylopectin, characterized by its numerous ramifications, does indeed present a very unique configuration; 12 glucose units are inserted on the main chain, which is also composed of 12 units of the same sugar. Alpha amylase creates three glucose units and sometimes more. Amylopectin, at-tacked by both amylases, protects itself against the damage that could be done to the bonds (1–6). Therefore, ramified residues remain, known by the name of "limit dextrins". Their hydrolysis is reserved for other enzymes specialized in the rupture of alpha (1–6) bridges. This is the case for pullulanase, ex-amylopectinase. As a result, the combined action of an amylase and a (1–6) glucanehydrolase causes a more extensive degradation of the amylopectin in glucose and maltose. These complexes are found in *Acetobacter*, enterobacte-ria, *Neisseria, Pneumococci, Bacillus*, etc. They are therefore rather wide-spread in the bacterial world. (See Figure 21.)

2.2.7. Pectins

With pectins, we are in the presence of gelatinous substances of which the vegetal world is the biggest producer. The prototype was discovered by Braconnot in 1825. In the unrefined state, this compound is formed by chains of pentosans and galactosans, the hydrolysis of which releases galacturonic acid and approximately 11% of methanol. The galacturonic acids are partially esterified by the methanol. Dimethyl-beta-methyl galacturanosides were iso-lated from methylated pectin. (See Figure 22.)

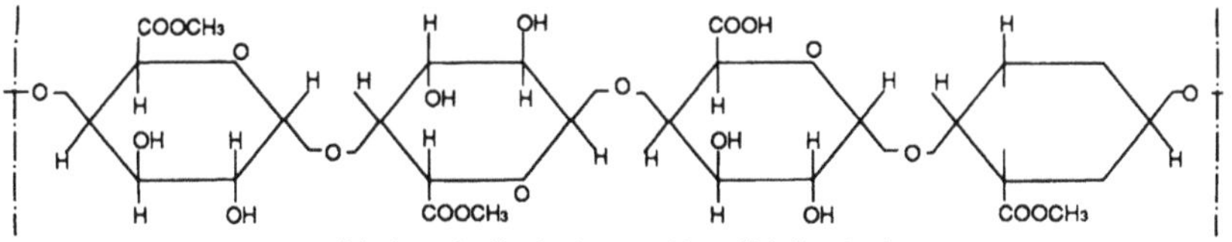

Polygalacturonic acid, carboxyl group, partially esterified with methanol

Figure 22. Pectins. In the raw state, they contain purified galactosane and pentosane, and they produce galacturonic acid and 11% methanol upon hydrolysis.

2.2.8. Glycogen

This polysaccharide is not only reserved for animal livers. It is indeed found in plants and a few bacteria. This vegetal or bacterial polysaccharide is nevertheless slightly different from the hepatic polysaccharide. In an acid medium, it produces glucose and, through enzymes, maltose. *Bacillus macerans* synthesizes amyloses close to starch in the presence of glycogen. The chains are naturally not as long, but they are much more rich in ramifications.

2.2.9. Nigeran (or Nigerose)

This polyglucosan was identified in 1914 in *Aspergillus niger* to which it owes its name. It was studied more specifically by Backer in 1953. Classified in the group of polyglucans, with Nigerose, it is characterized by an alternation of two types of glucosylic bonds alpha 1,3 and alpha 1,4. Apparently, not having attracted the attention of bacteriologists, it would appear more specific than *Aspergillus*. The same is true of another polyglucan, specific to *Penicillium*: Luteose, beta (1–6) malonic polyglucosan ester of luteic acid.

2.2.10. Alginates

These rather particular polysaccharides are made up of sequences of mannuronic acids or polyglucuronic acid. The exopolysaccharide of *Azotobacter vinelandii* and that of *Pseudomonas aeruginosa* provide a structure very close, if not identical, to the alginates of marine algae. They only differ in the proportion of C and the fact that the C_3 hydroxyl groups are most often acetylated.

At the beginning, it was thought that the sequences of bonds 1–4 of a copolymer of mannuronic-D-beta acid and of C-5 epimers of glucoronic-L-alpha acid alternated regularly. We now know, thanks to arranged enzymatic hydrolyses, that the sequences of monomers are randomized.

Although the bacterial alginates of *Acetobacter vinelandii* and *Pseudomonas aeruginosa* have the same basic structure as marine algae, the difference naturally lies in the acetylation of the hydroxyl groups. The content in *O*-acetyl

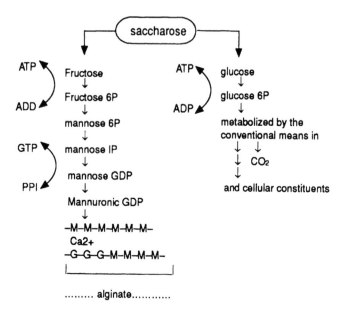

Figure 23.   Biosynthesis of alginate.

of the exopolysaccharides of *Pseudomonas* varies considerably depending on the strain, in a proportion of 0 to 13%. (See Figure 23.)

2.2.11. Madurose

This sugar, specific to Maduromycetes (Goodfellow[311]), is a 3-O-methyl D-galactose. The wall of these bacteria is structured with meso-2.6-diaminopimelic acid. This group of Maduromycetes includes eight genera, each one of which accepts several species — 26, for example, for the single genus *Actinomadura*. Madurose is therefore a sugar strictly specific to these bacteria.

It should be noted at this point that the classic agent of the "Madura foot disease" is now the leader of this genus along with the species *Actinomadura madurae*, the first specimen of which was described by Vincent in 1894! In any case, that is what contemporary systematicians have decided about it (*Bergey's Manual* Vol. 4, 1989, on Mr. and Mrs. Lechevalier's proposal).

3.   LIPOPOLYSACCHARIDES

Monosaccharides related to lipids have been the subject of essential studies since 1965. Jack and Strominger on the one hand, and P. Robbins et al.[122-123] on the other, discovered new types of carbohydrates, linked by monophosphate or diphosphate bridges to polyprenols, and long-chain lipids, the bactoprenol of which is a major representative. It participates in the biosynthesis, not only of liposaccharides, but also of peptidoglycans.

In the bacterial world, certain activators linked to lipids transport carbohydrates and even oligosides. They make them cross the cytoplasmic membrane whose wealth in lipids is known. The bacterial glycolipids are formed of a mannose-, rhamnose- and galactose-based trisaccharidic unit. Glycosylated nucleotides enter into the construction chain during the process. The role of isoprenoids was confirmed in *Enterobacter aerogenes* by Troy et al.[124] (1971) and Sutherland et al.[125] in other Gram-negative bacteria (1970). The sugars are managed, transformed in 6-phosphates, and then transformed in 1-phosphates according to the usual scenario. They are then transported by the monophosphate uridin on a pyrophosphate lipid. One of the transferases will carry out other operations to form the chains of lipopolysaccharides. Isopren intervenes in the form of isoprenoids in these biosyntheses. It is especially rich in the antigen O of certain enterobacteria, notably uropathogenic *Escherichia coli* (Svanborg-Eden et al.,[126] 1981). The production of this lipopolysaccharidic compound is considered fundamental on the level of the pathogenicity of the germ.

In general, bacteria have sufficient reserves in isoprenoids to ensure the simultaneous synthesis of peptidoglucans, polysaccharides, and lipopolysaccharides. The structural diagram of these lipopolysaccharides is summarized by a bond between the polysaccharides and a glycolipid.

The lipid is considered as a poly-*N*-acylglucosamine esterified by a phosphate of ethanolamine and fatty acids. In *E. coli*, chains of palmitic, hydroxymyristic, myristic, hydroxylauric, lauric, and acetic acids can be identified. Bacteria that are very far apart in the nomenclatura can present surprising relationships as far as lipopolysaccharides are concerned. Anders Sonesson and Erik Jantzen[127] (1992), for example, isolated polysaccharides of two *Legionella*. They shared a deoxyoctanic 2-ceto-3-acid. Moreover, the authors evidenced a common alditol and methylglucoside with two *Yersinia fredericksenii*. Such similarities are not exceptional.

Summary of a few bacterial polysaccharides:

Levans	Numerous *Pseudomonas, Alcaligenes viscosus, Aerobacter, Streptococcus salivarius.* Frequent in *Bacillus.*
Polymannans	*Bacillus polymyxa, Desulfovibrio desulfuricans*
Dextrans	*Acetobacter,* several streptococci, *Leuconostoc, Betabacterium*
Celluloses	*Acetobacter,* rather numerous *Agrobacterium*
Crown Gall	*Agrobacterium tumefaciens*
Starch — amylopectin	*E. coli, Corynebacterium diphteriae, Clostridium botulinum*
Glycogen	Numerous *Bacillus, Neisseriaceae,* a few *Pneumococci, Mycobacterium tuberculosis*

Alginates   *Pseudomonas aeruginosa, Azotobacter vinelandii*

4. COMMENT

Carbohydrates and polysaccharides represent the important basic elements of cellular structures. Due to this fact, they play a major role in the adherence mechanisms between the cells, irrespective of the organisms considered. New molecules have been discovered, resulting from the union between these carbohydrates or polysaccharides and other basic molecules such as proteins, lipids, and nucleoproteins. These compounds will be dealt with below.

5. PROTEINS AND GLYCOPROTEINS — COMPLEXES

Proteins, macromolecules universally widespread in the world of the living, result from the linking of amino acids. The most basic proteins release, after hydrolysis, only alpha-type amino acids. The others, often very complex, are created by amino acid and nonamino molecule combinations. The following two forms are distinguished: fibrous forms, insoluble in water, and globulous forms, soluble or miscible in acid, neutral, or alkaline media depending on their nature. Albumine and lactalbumine are common examples. Heat causes their denaturation. These proteins, generated by R-CH (NH$_2$) COOH amino alpha bonding, constitute the class of polypeptides on which carbohydrates are frequently grafted, creating glycopeptides.

Depending on their composition, proteins can be divided into two groups:

- Holoproteins, composed only of amino acids
- Very complex heteroproteins, formed of peptides on which organic or inorganic molecules are grafted, forming "prosthetic" groups

The molecular weights of these compounds are sometimes significant, since some of them exceed one million Da. The functional diversity of these complexes is also one of their features. They are found everywhere, but this monograph will limit their assessment and developments substantially. Each polypeptide contains at least 100 basic amino acids, and often more. Irrespective of the origin, each group contains the essential 20 amino acids. The stereochemistry is always three-dimensional. The structure is sometimes fibrous and sometimes helical. Disulfide bridges are observed, composed of cystein and ramifications. Size, shape, polarity of the ramifications, linking order of the amino acids, and ramifications of various substances are all singularities that provide each macromolecule with its specificity. In the globular forms, the chains are helical and wound. The genetically coded biosynthesis of these macromolecules is governed by enzymes.

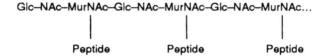

Figure 24a. First stage, peptide chain.

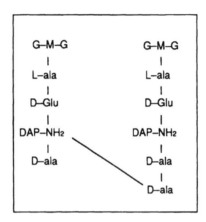

Figure 24b. Last stage, lateral peptide chain.

5.1. Polypeptides and Derivatives

5.1.1. Murein

The rigidity of the bacterial wall is ensured by a group of polypeptides, the main representative of which bears the name of "Murein" (from the Latin, *Murus* meaning wall or enclosure, and also meaning protection). This polypeptide is composed of muramic *N*-acetyl acid and *N*-acetyl glucosamine. The structural mode of the saccharide is close to that of cellulose with the difference being that the muramic acetyl contains a 3-0-D-lactyl group and that the average length of the chains remains limited.

More complex peptides formed of glutamic acid, D-alanine and sometimes L-alanine, lysine, or diaminopimelic acid are attached to acetylmuramic units. In *Staphylococcus aureus*, the tripeptidic chains are linked by a peptidoglycan bridge that will constitute the peptide wall. One of the specific constituents of bacteria is this diaminopimelic acid, a forerunner of a certain number of amino polypeptides and hetero-polypeptides. This acid itself can be transformed into a dipicolinic acid during the sporulation of bacteria having this faculty of resistance. The process is illustrated by a simplified diagram. The difference between the germs lies in the nature of the lateral chains and sequences and their bonding modes. The discovery of uridine dinucleotide in the wall of *S. aureus*, and of its inhibition by penicillin has improved our knowledge about

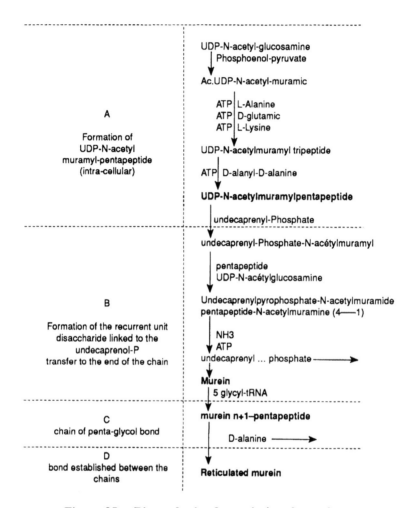

Figure 25.　Biosynthesis of murein in a bacterium.

mureins. The complex is represented schematically as shown in Figures 24–25, where N-Ac-Gluc, beta N-Acetylglucosamine and N-Ac-Mur, and beta N-acetyl muramic acid are present. The repetitive unit identified in *Micrococcus lysodeikticus* is a N-acetyl glucosamine–N-acetyl muramic acid with beta bonds (1–4) and (1–6). This bacteria, very sensitive to lysozyme, contains glycine at 1 mol/mol of glutamic acid. Some of these components are present in all bacteria. It is only a matter of proportions. If the sugars are determined in percentage of dried material weight, the values range between 50 and 80% in Gram-positive bacteria, compared to only 1 to 5% in Gram-negative bacteria. The components of streptococci reach 22%.

When all is said and done, the two amino sugars remain N-acetyl glucosamine and N-acetyl muramic acid. Alanine and glutamic acid are always present. The same is true for diaminopimelic acid, lysine, diaminobutyric 2-4 acid, and lastly L or D-ornithine. These reminders are only aimed at guiding the choice of labeled products (amino acids among others) for following the path of the bacteria in a given microenvironment. This is what we did on mollusks and bacteria (J. Brisou et al.,[128] 1982–1984).

Murein therefore surrounds the bacterial cells like a bag. As they enter into contact, they unite, thanks to these compounds which thus directly participate in adherence. They also present affinities with recognition lectins, due to carbohydrates. The pathogenic power is related to these structures, as will be shown in an upcoming chapter. In bacteria, one of the basic constituents of the wall is diaminopimelic acid, a forerunner of a certain number of polypeptides and heteropolypeptides. This acid can itself be converted, as we know, into a dipicolinic acid during sporulation. The simplified process can be represented as shown in Figures 26–27. The formula of muramic acid was specified in 1956 by Strange.[129,130] It is a 3-0-lactyl ether of D-galactosamine of which a few basic formulas are shown in Figures 28–29.

5.1.2. Sialic and Neuraminic Acids

Neuraminic acid is most often present in cellular membranes. Its role in the attachment of viruses and bacteria has been confirmed. This acid participates in the construction of gangliosides. The acetylated derivative is a result of the combination of a D-mannosamine and pyruvic acid. It is found in gangliosides and in glycolipids of cellular surfaces where it plays the role of specific receptor. All the derivatives of neuraminic acids belong to the class of sialic acids. These compounds include nine carbon atoms with a carboxyl group and a carbonyl and amino group.

All these complexes, including neuraminic acid, belong to cetons, the C_1 of which is oxidated on a carboxyl. Such a structure is opposed to alpha-type glucosidic bonds.

Sialic acids intervene actively in the agglutination phenomena and are associated with lectins. They are considered as terminal residues of oligo and polyosides belonging to gangliosides and ecto-cellular glycoproteins. They are born of glucosamine by following stages that can be summarized as follows:

Epimerization of N-acetyl glucosamine by an abandoned enzyme: acetylglucosamine phosphoisomerase, coded E.5.3.1.11.
The follow-up takes place according to the following process:
N-acetylglucosamine → N-acetylmannosamine.

In the presence of ATP, an N-acetylmannosamine 6P, united with phosphoenolpyruvate, will produce an acetylneuraminic phosphate acid and lastly the acid itself, which constitutes the source substrate for other complexes, thanks to the sialotransferase activity. Sialic derivatives, all belonging to the N-acylate of neuraminic acid found on the cellular interfaces, actively participate

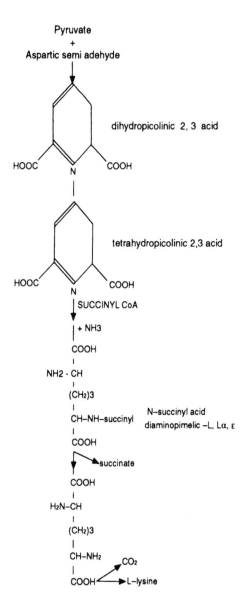

Figure 26.   **Dipicolinic and diaminopimelic acids characteristic of bacteria biosynthesis process.**

in the adherence of certain microorganisms, notably chlamydial infections and some strains of *E. coli* hungry for mucous membranes. The most striking derivatives are sialic-pyruvic and sialic-lactic acids and one of these polypeptides is known by the name of Corinic. These acids were closely studied on

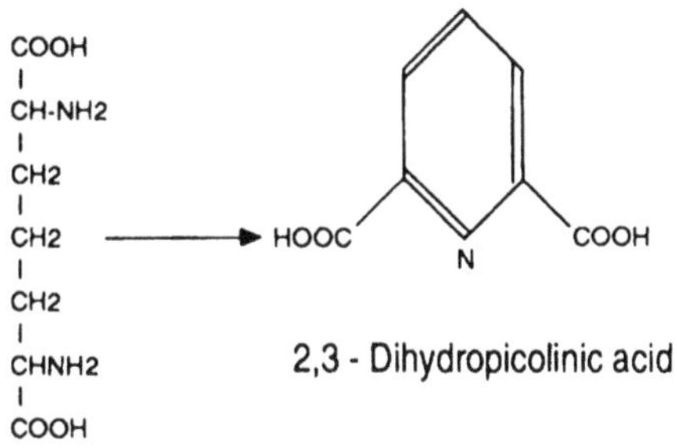

$$COOH$$
$$|$$
$$CH\text{-}NH2$$
$$|$$
$$CH2$$
$$|$$
$$CH2 \longrightarrow HOOC \quad \text{(ring with N)} \quad COOH$$
$$|$$
$$CH2$$
$$|$$
$$CHNH2 \qquad \text{2,3 - Dihydropicolinic acid}$$
$$|$$
$$COOH$$

Figure 27. Diaminopimelic acid.

erythrocytal membranes. Sialidase or neuraminidase (EC.3.2.1.18) hydrolyzes the cellular coating. It modifies the adherence capacities and possibilities of the microorganisms and even parasites considerably, hence the medical interest of these experimentations. The two most common sialic acids are N-acetylneuraminic and N-acetylglucosaminic acids.

In 1983, Gallacher et al.[131] studied the relations established between the sialic acids of the cellular interfaces and agglutination by wheat germ lectins. The specific site of hemagglutination is constituted of three subunits, each one of which is complementary to N-acetylglucosamine. The free sugar behaves like a weak glucoconjugate inhibitor. However, the di- and trisaccharides such as chitobiosis and chitoriosis act vigorously. This suggests that the affinity between the complexes is conditioned by agents acting simultaneously with two or several available subunits. Neuraminidase inhibits agglutination. Sialic acid, sensitive to the enzyme, therefore plays an important role in the agglutination process.

The wheat germ lectins recognize both sialic acid and the Glc-Ac residues. This acid, placed at the end of the polysaccharide chains, is undoubtedly preferentially localized. Spatial and stereochemical constraints are taken into account. A specific sialic acid was identified in the fimbriae of three *E. coli, Neisseria gonorrhoeae, Pseudomonas aeruginosa* and in the fibrillae of *S. sanguis* (Buchanan et al.[135] in 1978, Murray et al.[133] in 1982, and Rampal et al.[134] in 1983, and summarized by Lindahl et al.[132] in 1982–83.) These bacterial lectins, located on the fimbriae, cause agglutinations by binding with another molecule, glycophorine, of which the affinity for sialic acids is established in the O-acetyl groups in C_4 and N-Glucosyl in C_5. This specificity can play a role in the pathogenicity of the germ. The penultimate sugar of the molecule can increase the activity of the system, notably if it is a N-acetyl-neuraminic residue, depending on its positions on one of the nine carbons:

$$R1 = COOH, R2 = OH, R4 = H, R5 = COCH_3.$$

The links between chemical structure and pathogenic power are obvious.

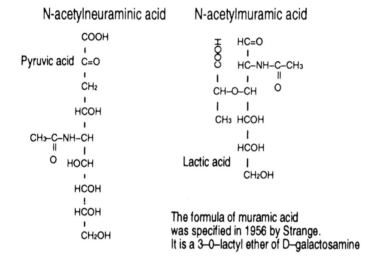

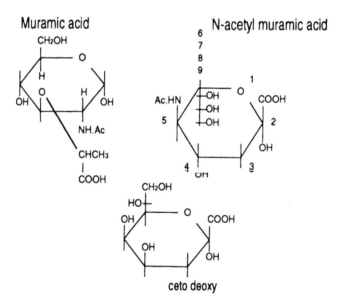

Figure 28.   Basic formulas.

5.1.3.   Glycoproteins — Lectins — Pilins

A certain number of specific glycoproteins were localized on structural lectins which are fimbriae and fibrillae. Their role in adherence has not been questioned since the studies by Wadstrom et al.[136] (1981). The type 1 fimbriae are attached to mannose or to the residues of this same sugar. On these

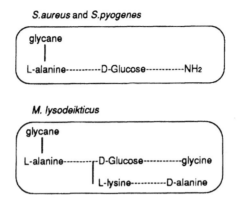

Figure 29. Schematic structure of mucoproteins.

filaments, notably in fimbriae, numerous binding sites have been localized. Most are destroyed by the ultrasounds that shorten the filaments. The zones of affinity are therefore spread along the entire length of the organella and not only on the ends.

There are notable antigenic differences in the structures of all these bacterial fimbriae. The contact specificity with mannose has been confirmed in *E. coli, V. cholerae, Klebsiella, Aeromoonas hydrophila*, etc. Other bacteria prefer fucose. D-galactose, *N*-acetyl glucosamine, and sialic acid retain other species, notably *Mycoplasma pneumoniae, Neisseria gonorrhoeae, Bordetella pertussis*. These aspects will be dealt with again in the chapter on cellular receptors. They also affect pathogenicity.

5.1.4. Uronic Acids

The participation of uronic acids and their derivatives is generally recognized in the structure of Gram-positive bacteria. They are abundant in *Pneumococci* and *staphylococci*. The basic elements are *N*-acetylglucosamine, glucuronic acid, and mannoglucuronic acid in the pyranoside form. Rhamnose, galactose, cellobiouronic acid, and galacturonic acid should also be added. The bonds are ensured in alpha or beta on the modes: (1–3), (1–4) and (1–6), each endowing the molecule with its specificity. These polysaccharides are synthesized based on nucleotidic sugars, UDP-glucose, UDP glucuronic, and UDP glucosamine, usual forerunners of all these syntheses. These complexes are gifted with strong antigenicity and take a substantial part in the construction of capsules. Polysaccharide layers of the same type are also present in the walls, between the protein layer and the mucopeptidic layer. The viscous capsular material of A and C streptococci are examples.

5.1.5. Hyaluronic Acid

This heteropolysaccharide, one of the most important, is composed of glucuronic acid alternating with N-acetylglucosamine. This structure makes an excellent lubricant, a cohesion substance of the cells, and a capital agent of adherence. That is the reason why we always used hyaluronidases in the releases. This acid falls into the group of glucosaminoglucans, but its biosynthesis follows different paths, on this level drawing it closer to sulfate keratans, sulfate chondroitins, dermatans, and even heparin. This particularity led biochemists to the comprehension of a new synthesis mechanism unknown until now. It occurs on the cytoplasmic membrane and appears, in some cases, to be induced by retinol.

This fact was confirmed due to the use of basic-labeled products. Once the young chains are organized, it is not long before they are exteriorized. The process is inhibited by detergents and thiols, but is activated by phosphates. A certain number of bacteria and other microorganisms synthesize hyaluronic acid which is abundant in capsulated Gram-positive germs. A relation was recognized between the presence of this acid and virulence. Strains previously treated with hyaluronidase lose a large part of their pathogenic power (10 times less virulent). In streptococci C, the enzyme destroys hyaluronic acid and decreases the virulence of the germ by 10^5. As early as 1937, Kendall[137] had already isolated the hyaluronic acid of hemolytic streptococci.

5.1.6. Chitin

This polypeptide is a Poly-2-N-acetamido-2-deoxyglucan. It is encountered in a certain number of microorganisms and in lower fungi. It is an osamin widely spread throughout the living world. Bacterial chitin resembles that of crustaceans. It can therefore be supposed that it is constructed by following the same pathways:

A Poly-2-N-acetamido-2-deoxy-D-glucosa linked in glucosidic beta-(1–4) and in a linear chain. There are no lateral chains.

5.1.7. Teichoic Acids

Breakdown of teichoic acids in a few bacteria:

Polymer Type	Glycerol	Ribitol
Lactobacillus arabinosus	–	–
Lactobacillus casei	+	–
Staphylococcus aureus	Traces	+
Staphylococcus aureus Oxford	+	+
Staphylococcus citreus	+	–
Staphylococcus albus	+	–
Bacillus subtilis	–	+
E. coli	Traces	–
Corynebacterium xerosis	+	–
Streptococcus faecalis	+	+

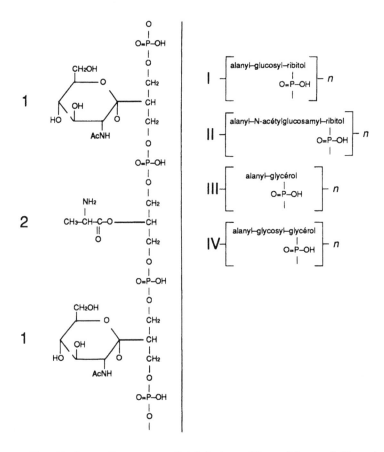

Figure 30. Left — Sequence of teichoic acid, residues of *N*-acetyl-glucosamine (1) and *D*-alanine. Right — Four diagrams materializing the main types of teichoic acids.

These complexes (see Figure 30) were discovered in 1968 by Baddiley,[138,318] Buchanan, and Latters in *Lactobacillus arabinosus*. Since then, a certain number of similar compounds have been evidenced, notably polyol-phosphates which are very widespread in Gram-positive species. These acids were extracted from *Bacillus subtilis*, various lactobacilli, staphylococci, and streptococci. In 1983, Whigham and Kleinman[317] studied these complexes in *B. globigii* and *S. aureus*. They concern immunology due to their antigenic structures, and they are attibuted with an anticytotoxic action.

Certain capsular polysaccharides contain phosphate ribitol. Such is the case of *Pneumococcus IV*, the hydrolysis of which releases phosphate ribitol, galactose, glucose, and rhamnose. The types of bonds differ from those that are usually observed. These substances are extractable by acids diluted at 10% at 4°C after several days of contact. Linked to murein, they participate in approximately 50% of the parietal structures. This proportion confirms their value.

Such compounds are, as a whole, considered as polymers of phosphate ribitol and various sugars in which glucose and galactose, joined by alanine, dominate. Certain teichuronic complexes, anionic polymers, free of PO_4, owe their originality to the presence of acid sugars in their molecule. Their synthesis occurs following PO_4 depletion.

The ester D-alanine bonds are fragile in an alkaline medium. *Streptococcus faecalis* delivers a polyglycerophosphate. These teichoic acids, notably those extracted from *B. subtilis*, various lactobacilli, *S. aureus*, type A hemolytic *Streptococcus pyogenes* and other bacteria, have been recognized for over 30 years. The position they occupy in the parietal architecture of certain bacteria causes them to intervene in the adherence process. *S. faecalis* is endowed with a phosphate ribitol complex containing N-acetylglucosamine and glucose. The isolated product of *Bacillus megaterium* contains glycerol. The teichoic acids are united with other parietal materials by salts and hydrogen bridges. In 1961, Mandelstam et al.[139] estimated that the teichoic acid would be, in certain circumstances, linked to the glycopeptide by a peptidic bridge thrown between D-alanine and the acid itself, creating a genuine ester.

In any case, these complexes participate actively in the rigidity of the bacterial walls. They are, in principle, specific to Gram-positive bacteria; nevertheless, some ribitol-based waxes, phosphates, glucoses and mucopeptides were isolated from *E. coli* by Lilly[140] in 1962. The acid isolated from *B. megaterium* differs from those usually isolated from Gram-positive bacteria. Studies conducted using Carbon 14 show that it contains anhydroribitol. These teichoic acids are localized on the surface of the wall. They thus take part in the antigenicity and are likely to facilitate the passage of ions through the walls. According to Archibald et al.[141] (1976), they behave like "regulators".

In addition to these superficial components, there is another intracellular group where glycero-teichoic acids dominate, localized between the cytoplasmic membrane and the inner side of the wall (Hay et al.[142] 1963). In *S. aureus*, the importance of magnesium is noted in the biosynthesis of these complexes. In a medium limited in PO_4, the parietal compound is no longer reproduced. A new, nonphosphorated, molecule comes to life in the composition of which N-acetylglucosamine and N-acetylaminoglucuronic acid are identified. The teichoic acid of *S. aureus* also proves to be an excellent antigen. These composition variations of the walls involve the mechanisms of adherence and especially contribute fundamental data applicable to the preparations of vaccinations. This will be dealt with again at the end of this monograph.

The wall of *M. lysodeikticus* contains polymers of glucose and phosphorated N-acetylmuramic acid (Perkins[143] in 1963). According to Salton[144] (1953), there appears to be from 0.09 to 0.13% of PO_4 in the complex. This is also Ghuysen's[145] opinion (1968). *S. aureus* only cultivates well on rich media. Moreover, it requires the presence of Mg^{++} and PO_4. In media limited in phosphorus, at 35°C and with a pH level of seven, the synthesized polymer contains glucose and mannuronic acid. In the presence of a limited dose of Mg, there are 3.1% phosphates. It is thus possible to extract a teichoic acid containing a (1–3) polyglycerol phosphate substituted in position $1\text{-}C_2$ of glycerol by N-acetylglucosamine. It can therefore be observed in the Gram-positive bacteria that the chemical composition of the walls is extremely variable. Certain authors, including Ellwood and Tempost, consider that all these variations are subject to genetics, a point of view which currently seems obvious.

The frequency of the genetic exchanges and mutations between the bacteria is known. They cause selections in function with the media, microenvironment and living conditions. As a result, the germs, irrespective of their type, do not mandatorily or indifferently attach to a given interface.

5.1.8. Summary of the Biosynthesis of Peptidoglycans

This biosynthesis takes place in three stages:

1. Birth of a nucleotide: UD, muramylpentapeptide.
2. Half of this nucleotide is transferred on a phosphorylated lipid. An Ac-glucosamine is added to the muramyl peptide that generates the sequence: Lipid-P-P-muramylglucosamine-pentapeptide.
3. Transfer of the membrane towards the wall of this complex on the peptido-glycan produces peptidoglycan $\rightarrow$ Lipid-P-P + peptidoglycan-glucosamine-muramyl-pentapeptide on this wall.

On the wall, a peptidic bond is made between the new unit and an already present chain, with loss of an alanine molecule (transpeptidation) according to Figures 24a and b. There is no ATP input and the beta lactamines (penicillin and cephalosporins, for example) inhibit transpeptidation.

5.2. Lipoproteins and Derivative Complexes

5.2.1. Lipoteichoic Acid

These lipidic derivatives of teichoic acids were first studied in type A hemolytic streptococci. It is recognized that these products, like the previous ones, participate actively in adherence.

The first studies date back to 1972–1975. We owe a substantial part to Toon et al.[319] as well as to Ganfield and Pieringer. The research was continued by taking S. faecalis as a reference subject. It was established that 28 to 35 units of glycerol-phosphate take part in the construction of the complex. Partial substitutions are noted with kojibiosyldiglyceride residues or more often with a phosphatidyl-kojibiodiglyceride residue.

It should be specified that kojic acid is a 2-hydro-methyl-5-hydroxy gamma pyran. (See Figure 31.) Certain bacteria cultivated in the presence of glycerol or dulcitol elaborate this acid with the configuration of a gamma pyran. (See Figure 32.) The bonding mode between teichoic acid and the lipid chooses the terminal phosphate of the acid to attach it on the C3 of external glucose, or

CH2OR'
OH
+–OR'
CH2–O=P–O–CH2
O
O
CH2OH
CH2OR'
+–OR'
O
OH
HO
O–CH2
CH2
28-35
RO+H
OH
CH2
O
HO
O

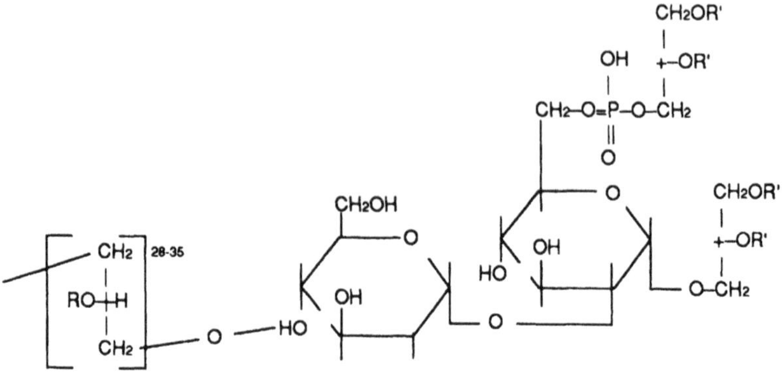

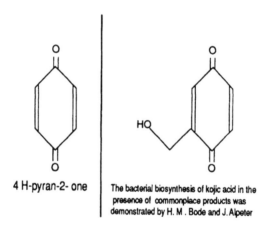

Figure 31. Structure close to lipoteichoic acid of *S. Faecalis*. R = kojibiosyl residue, and R′ = fatty acid. (Data from Toon et al.,[319] and Ganfield and Pieringer.[320])

O
O
HO
O
O
4 H-pyran-2- one
The bacterial biosynthesis of kojic acid in the presence of commonplace products was demonstrated by H. M . Bode and J. Alpeter

Figure 32. Left — Pyran. Right — Kojic acid, a derivative of pyran.

sometimes on carbons C3 or C4 of internal hexose of the lipidic complex. Toon[319] (1972), and Ganfield and Pieringer[320] (1975) suggested a structure in which (R) corresponds to a kojibiosyl residue and (R') to the fatty acid.

In *S. aureus,* the lipidic unit appears to be a gentiosyldiglyceride, with traces of *N*-acetyl glucosamine. Studies done with labeled elements, notably 32P-glycerol, indicate that the bond between the teichoic acid and the glycolipid are located on the level of the P terminal of the chain and on the free 6 of the glycolipid.

In 1975, Wicken and Knox[146] suggested a schema of lipoteichoic acid in which the terminal glycerol of the chain or a substitute would be acetylated. The terminal lipidic residue corresponds to a glycolipid of the cytoplasmic membrane. For Shockman[147] and Slade[148] (1964), who studied a strain of *S. faecalis*, the polymers are only "discreetly" associated with the said membrane. For Duckworth[149] (1977), the hydrophobic poles of the molecule react with the double lipidic layer of the membrane, whereas the hydrophilic pole of the teichoic chain, localized on the outer surface, penetrates in the zone located between the membrane and the bacterial wall. Wicken and Knox[146] (1975) had already shown that the lipidic fraction of the lipidoteichoic molecule is truly buried in the membrane. Another portion crosses the layer of peptidoglucans and emerges at the surface where it enters into contact with the hydric environment. Beachey et al.[150] (1980) consider that in such conditions the lipoteichoic acid forms stable complexes with some of the parietal proteins which can serve as models.

In summary, these complexes establish a genuine network of lipoteichoic acids and proteins at the surface of the bacteria to constitute the coating that forms the fibrillae, which are attached like sexual pili on the cytoplasmic membrane. The union of these complexes is such that the lipidic extremities of the teichoic molecules can "float" on the surface, move, and react directly with the components of the eukaryote cells encountered. This phenomenon leads us into the field of adherence.

Group A streptococci directly secrete lipoteichoic acid in favorable culture media. This "protonized" acid facilitates hydrogen bonds with surface proteins. Lipoteichoic acid will thus be truly "anchored" on the cellular surface. Beachey and Ofek[151] (1976) specified that the M proteins of the streptococci are in relation with the teichoic acids associated with the fibrillae. The ratios between the lipidic fraction of the lipoteichoic acid of *B. licheniformisi* and the membrane glycolipids were confirmed by Button et al.[152] (1976). There is a similarity between the composition in fatty acids of the lipoteichoic acid, but the same is not true for all bacteria.

Esterifications are carried out, in all cases, on the level of the glycerol residues. It is probable that the lipoteichoic acid and its so-called "carrier" structures have different identities, as all the studies made on this subject concerning Gram-positive bacteria imply. On the other hand, it seems that the antigenicity of the lipoteichoic complexes is limited to a few bacterial genera, notably to streptococci and lactobacilli. These acids, the importance of which is understood, offer a wide variety of complex compositions for analysis. Certain unsaturated fatty acids dominate the whole. The main ones to be retained are, in order of carbon content, myristic acid in C14, palmitic acid in C16, and stearic acid in C18.

5.2.2. *Glucido-Lipido-Polypeptidic Complexes*

Lipoproteins are, in the end, associations of polar lipids and proteins which often serve as sensors and transporters. They can sometimes become complicated by receiving carbohydrates to yield glucido-lipido-polypeptidic units, the foundations of certain endotoxins of which the O antigens of enterobacteria are the most common examples. This data will only be recalled for memory since these endotoxins do not directly concern adherence but rather the penetration mechanisms of bacteria in the tissues and their pathogenic power, once adherence has been achieved.

The endotoxin of a certain number of Gram-negative bacteria, *Salmonella typhi*, is linked to the O antigen, the constitution of which has been recognized since the studies by Westphal and Luderitz. This endotoxin is a glucido-lipido-protein complex resulting from the association of a polyoside, a lipid (A) and a protein to which another nontoxic, heterogenous lipid (B) is added. The glucidic fraction is represented by a phosphorylated polyoside. The protein varies depending on whether the bacteria is in phase S (smooth) or in phase R (rough). The lipids (A) correspond to a heterogenous phospholipid: the association of fatty acids, phosphates, and lipoproteins as well as a glucosamine. This fraction is responsible for pyrogenesis.

6. PARTICULARITIES OF MYCOBACTERIA

Mycobacteria belong to Section 16 of "Bergey's Manual" which accepts four kinds: *Mycobacterium, Nocardia, Rhodococcus* and *Corynebacterium*. The common factor of these bacteria is the presence of mycolic acids on their walls. These acids are noteworthy in the mechanisms of adherence and in the behavior of these microorganisms. Certain are difficult to culture, or even for the time being, impossible, as testified by *M. leprae* (described by Hansen[153,154] in 1874) and resistant to all culture attempts. Current studies show the extreme complexity of its structure, as well as that of other species.

A polysaccharide deprived of glycogen was extracted from *M. tuberculosis hominis* and *bovis*. This complex contributes to antigenicity. Its analysis provides glucose units and a disaccharide. It resists the attack of the beta-amylase and appears linked to lipidic groups. The second sugar is a D-glucosamine. The polysaccharide is, in the end, composed of a D-glucopyranosic chain constructed on the alpha (1–2) mode with a molecular weight close to 4.9 $\times 10^5$ Da. The peptido-glycolipidic complexes contain 2 mol of meso-diaminopimelic acid and 2 mol of L-alanine and 1 mol of D-alanine and glutamic acid (Asselineau et al.,[155-157] 1958).

As far as *M. tuberculosis* is concerned, the mycoside of *M. avium*, in its fraction C, contains D-leucine (Snell et al.[158] in 1955). These mycosides are glycolipids of which two types are distinguished:

- the (A) in one half, includes a di- or tri-mycocerate of aromatic alcohol,
- the (B) is characterized by two molecules of a ramified chain acid fraction linked to a methylated triolphenolic (Lederer,[159] 1961).

The content in lipids of the wall of certain mycobacteria can reach 64% (*M. tuberculosis*).

A particularity of Mycobacteria lies in the presence of waxes, esters of monobasic acids with long chains of carbon and monohydric alcohols. These complexes are also carriers of free fatty acids, alcohols, and sometimes hydrocarbons. The long chains free of glycerol are differentiated in this way from triglycerides. The bacteria responsible for tuberculosis are rich in diesters of phtiocerols. The biosynthesis builds compounds containing up to 90 carbons, and follows, as a whole, a march close to the one that organizes the lipids.

The vegetal waxes undoubtedly play a role in the adherence of microorganisms on the surface of plants. Certain marine organisms, notably zooplankton, also have waxes (Benson et al., 1972). The synthesis precursors are, above all, glucose and alanine, incorporated in an alcohol half (Hendersen and Sarget in 1980). The fatty alcohols result from the association of the sugar whereas the acid fraction of the wax is of food origin.

Here are a few examples:

Palmitic acid	C16 = hexodecanoic	$CH_3-(CH_2)14-COOH$
Cerotic acid	C25 = pentacosanoic	$CH_3-(CH_2)23-COOH$
Myrycic acid	C30 = triacontanoic	$CH_3-(CH_2)28-COOH$
Hexacosanol alcohol ceryl		$CH_3-(CH_2)24-CH_2OH$

The mycolic acids are quite varied as Asselineau, on the one hand, and A.H. Ross,[160] on the other hand, confirmed. These acids are found in *Nocardia, Rhodococcus*, and *Corynebacteria*. The number of carbons in these compounds ranges from 22 to 90%. Peptidoglycans containing muramic acid participate in the structure of the walls. Among the dominant sugars, there are arabinose and galactose, with the most striking amino acids such as alanine, glucosamine, and glutamic acid.

Phtiocol, an important molecule, is a 2-methyl-3-hydroxy-1:4-naphtoquinone resulting from the alkaline hydrolysis of the lipidic fraction of *M. tuberculosis*. Mycomycine, one of the fractions responsible for the bacteria, results from the esterification of two molecules of mycolic acid by a 6-hydroxytrehalose.

Mycobacterium leprae

This bacteria, which cannot be cultivated in the current technology, is inoculable in the armadillo and in a race of hairless mice. It was thus possible to collect a sufficient amount of material to continue the chemical study of the structures. Rastogi et al. (1984) confirmed the multistratified aspect of the walls and the properties of various layers. Freeze-etching of the wall separates two sheets, a relatively smooth outer surface and a filamentous surrounding surface, which is itself a mixture of tubules and lamella sprinkled with particles. A genuine skeleton of peptidoglycans protects the cell. This defense wall

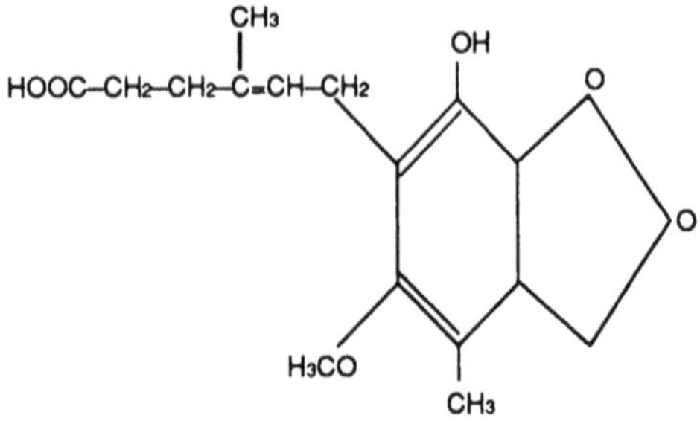

Figure 33.   Mycophenolic acid.

is built of repetitive units of N-acetylglucosamine and muramic glycosyl acid, linked by tetrapeptides. Moreover, glycine and alanine are detected in them. An association of polysaccharide lateral chains of arabino-galactose, esterified by mycolic acid, is grafted on the muramic acid. Lipopolysaccharides, mycolic acids, and two long hydryphobic chains of 60 to 90 carbons pile up in these envelopes. Lastly, the count increases with diesters of mycocerotic acid and peptidolycolipids.

According to Draper,[161] there is a mycosic capsule corresponding to the mycosidic tegument. In 1987, Gaylor and Brennan[162] identified mycosides A, of the class of oligo-glycosylphenolic-phtiocerol diesters, combined with sugars, such as 3.6-di-0-methylglucose, 2.3-di-0-methylmannose, and 3-0-methylrhamnose. There are probably still nonphenolated myco-cerosic acids, nonglycosylated phtiococerol dimycocerosic acids, lipo-arabino-mannans, and lipomannans. The aliphatic chains combine with the mycolic acids.

Minnikin[163] suggested a molecular model of the envelope of mycobacteria (1987) on which the double lamellar layer with long chains of mycolic acid is distinguished and linked to hydrophobic alignments of phtiocolmycocerosate, which are always present in Hansen's bacillus. It is probable that the sulfolipids unite with mycolic acid to constitute the double layer.

It is a genuine armor-plate that protects $M.\ leprae$. It is composed of a microcapsule made up of two symmetric sheets rich in polysaccharides and mycolic acids with which cerosides and phenolic glycolipids join, and lastly, tuberculostearic acid. A phenolic glycolipid provides a specific serological response. (See Figure 33.) This whole unit endows this bacteria with an out of the ordinary specificity that could be one of the causes of all the cultivation failures for more than a century!

7.   COMPOSITION DIFFERENCES OF THE BACTERIAL WALLS (CELL WALLS)

The composition differences of the bacterial walls are significant as indicated by what has just been developed. All the components are genetically programmed, and the life of the cells is closely subjected to the conditions of the environment. Such is therefore the case for their behavior with respect to interfaces. In Gram-positive bacteria, the wall only has one layer.

The components are

- Peptidoglycans
- Teichoic acids
- Lipoteichoic acids
- Polysaccharides
- Protein M (in *Streptococci*).

In Gram-negative bacteria, the wall has two layers:

- One thin inner layer (peptidoglycans)
- One thicker, outer layer (lipoproteins, lipopolysaccharides, proteins)

The Gram-positive bacteria are the only ones to have teichoic acids. The lipoproteins and lipopolysaccharides are only present in Gram-negative bacteria. The peptidoglycans are uniformly distributed in the walls of Gram-positive germs, whereas they are localized in the inner layer in Gram-negative bacteria.

8. CONCLUSION

This concludes Part I which is devoted to the structures and chemical composition of the elements taking part in the adherence of bacteria on all the interfaces, their invasive power, and their pathogenicity. The techniques that will be developed in Part III are based on this data.

PART II
Bacteria Life in the Wild

1 ADHERENCE IN LIVING TISSUES

He who wants to hunt bacteria and attempt to "release" them must know where and how they are hidden...

1. CELLULAR RECEPTORS

The whole biosphere is populated with prokaryote microorganisms such as bacteria or viruses, microeukaryotes represented by yeasts and lower fungi, and Protozoa. The number of bacteria that we house surpasses that of the cells making up our bodies. Part II presents a brief report of microbisms in living beings and microbiocenoses spread throughout the environment.

To attach themselves, bacteria must find, while equipped with all the adherence organs reviewed in Part I, host sites on living interfaces. These host sites are called "cellular receptors" and are as complex in animals as in vegetals. Cellular structures have been evidenced by electron microscopy and freeze-etching. Their compositions, better known because of immunology and biochemistry, are also extremely complex. The cytoplasmic membranes are, like in bacteria, covered with microfilaments and glycocalyx. Microtubules are very substantial here. They follow the tribulations of mitoses and undergo shape and size modifications. In 1984, K. Weber and M. Osborn[164] described a "cytoskeleton" rich in filaments which will be discussed later.

One of the chemical components of these cytoplasmic structures is known by the name of Actin and is a basic protein of these microfilaments. It is made up of fibrous chains on which other proteins are attached. These unions generate a variety of fibrous or gelified complexes. Other cellular substances have the job of picking up the ions and ensuring their circulation. The cells themselves can only be organized in tissues because of mechanisms ensuring their adherence. Among these compounds, there is sialic acid along with many other cell-adhesion molecules (CAM), including those studied by G. Edelman.[165] Although specific receptors such as glycocalyx are implanted on the majority of epithelia, other molecules or organella play the role of sensors and fixators in the same way. For example, sialiglycoproteins occupy the surface of human erythrocytes for which mycoplasms have a distinct affinity. A glucoside receives *Escherichia coli* specifically on the urinary epithelium (Svanborg-Eden et al.[166]).

Lectins constitute a major group of cellular receptors. They too are universally spread in all fields (Part I, Chapter 1: Adhesins). The specificity of these receptors has been confirmed by numerous observations. For example, some lactobacilli in rats only attach themselves to intestinal cells in this animal. Other concrete examples will be given later. All microorganisms populate cutaneous and mucous coverings. They proliferate in the digestive tract of all animals. They participate in the biodegradation and transformation processes of food and synthesize vitamins, etc.

Alongside these germs which are essential for vital balances (or inoffensive commensals), there are pathogenic germs that will also be taken into account. Bacteria attachment on living surfaces takes place in several stages.

1. Chemotactic actions related to adsorption promote the joining of bacteria and tissues. Some epithelia are rich in mucus that promotes the capture of microbes. The concept of free energy on which Buscher et al.[167] (1983) insisted should be taken into account here by studying the formation mechanisms of dental plaque. The effects of this free energy were noted *in vitro* and *in vivo*. Results obviously varied depending on the materials used.

2. Bacteria then penetrate the mucous layer.

3. Reversible adhesion will be conditioned by the nature of the receptors in the mucous gel or on the native microbism, which is associated with the tissue as is the case in the intestine or other mucosas. If everything takes place normally, the last stage will be the final adherence of the bacteria on the cell surface. The installed germs will find a shelter and possibilities for reproduction. If they are pathogenic, they will colonize the tissues and provoke characteristic lesions.

4. In the end, the sequence: chemotactic attraction of the bacteria on the surface of mucous gels → penetration in the mucus → adhesion and then possibly adherence on the specific receptors, summarizes the usual scenario of attachment on a tissue. The association with the mucosa is an essential condition, as noted by R. Freter[168] in 1981. However, as adhesion is reversible, bacteria can, in certain cases, be rejected, as will be shown by a few examples.

Confirmation of this installation was established by Le Brec et al.[169] (1965–66) thanks to histological methods, then by Dubos et al.,[170] Hoffman, and Frank. Specific actions coordinate the relations between the bacteria. nature, or the specific properties of the mucous surfaces. Adhesins, previously

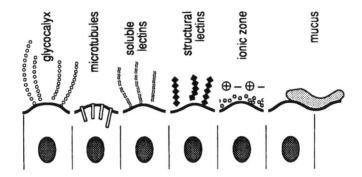

Figure 34. Relation between adhesins and cellular receptors.

studied (Part I, Chapter 1), were defined in this period as being responsible for the relations between germs and pathogenicity. These studies were continued in 1979 by Sugarnam and Donta,[171] as well as by Thgorne et al., and Svanborg-Eden and Jodal.[166] (See Figure 34.)

A mucous surface includes a glycoprotein material where glycocalyx and another group of glycoprotein substances (different from the former) are attached. These substances, elaborated by specialized cells, cover the epithelium with a gel that certain authors have compared to protective "battlements". Nevertheless, certain facts tend to partially refute this interpretation. Inert particles and bacteria, whether mobile or immobile, do indeed cross this gelatinous layer easily. Studies by Allweis[172] (1977) show that mobile bacteria passing through the area surrounding the mucosa are attracted by a chemotactic gradient. Finalist's conceptions attempt to confirm these interpretations. For example, they suggest the expression "tactic gradient" inherent to the mucosa and "the aptitude of the bacteria" to respond to the invitation of these taxes.

Analogous facts are observed in phytopathology. Some parasites seem to adapt to specific cellular structures and to the presence of carbohydrates. Experiments performed *in vitro* on the mobility of bacteria, their power to synthesize flagella, and the receptors capability of responding to taxins, are not transposable to the conditions in "the wild". This biosynthesis is only possible *in situ* if a certain number of chemical and nutritive conditions are present in a microenvironment, which is not obvious *in vivo*.

These same comments naturally apply to the progress of the second stage, concerning the penetration of the bacteria in the layer of mucus-gel. It is suggested that the adhesins react with receptors present on this level. Rolf Freter[173] reported observations made on *Vibrio cholerae*, which adhere strongly to rabbit intestine fragments. The last stage clearly corresponds to a state of irreversible adherence. The highly instructive experiment conducted by a team of researchers from the Ciba-Geigy laboratories in Bale (Copen et al., 1980) monitored the attachment process of an *E. coli:* SS-142 strain, originating from

Bristol where it had been isolated from a patient's urine. In the presence of a monocellular layered culture of intestinal epithelium, the germs observed, labeled with 14C, show that more than 98% of the radioactivity measured is related to bacterial bodies. The percentage of $E.$ $coli$ adherent to the cells becomes practically constant when the microbial concentration is maintained between 5×10^6 and 1×10^9 germs/ml. It is thus suggested that adherence is more related to microbial concentration than to volume. This adherence follows the process previously traced: an initial, short, and reversible stage, corresponding to adhesion, soon followed by an irreversible state characterizing adherence. Here, the authors developed an analogy between the dynamics of adhesion and enzymatic kinetics governed by the laws of Michaelis-Menten (Part I, Chapter 1). The essential elements of the phenomenon remain the concentration of bacteria (B) and the number of cellular receptors (R).

For concentrations of (B) ranging from 1.35×10^8 to 4.2×10^9/ml, at 37°C, the adhesion of the bacteria to the cells reaches saturation in 10 to 20 min. The attachment in two stages and the consistency of the reaction speed are confirmed at least within the limits of these 20 min. The first, reversible stage is interpreted as a "recognition" action. During this research, it was noted that raffinose inhibited the adherence of the experimented strain. Bactrim acts in the same way. These kinetics are applicable to many other microorganisms but they concern experiments conducted in $vitro$ on cultures of selected cells and "tamed" bacteria. It would be careless to transpose these "facts" to organisms in $vivo$ in a natural medium and conditions.

However, on a fundamental level, these laboratory observations are instructive. They demonstrate that bacteria, capable of settling definitely on tissues, present different characters of nonadherent germs. Overall, they appear less sensitive to antibiotics (Gwynn, 1981)[174] and to defense systems of the host. The mucogel layer covering all the epithelia is relatively easy to destroy by hydrolases.

In 1971, by using papain, Forstman confirmed what we had established with R. Babin, A. Rigaud et al.[175-178] in 1949, 1952, 1956, and again in 1975. At the time, we had insisted on the therapeutic value of this vegetal protease combined with antibiotics (penicillin, bacitracine, streptomycine). The enzyme offers the major advantage of reacting to physiological pH and to the normal temperature of the human body. It rapidly dissolves the necrosed tissues and puses rapidly which are microbe substrates. It strips infected wounds, liquifies the pus from furuncles, releases bacteria, and sensitizes them to the action of antibiotics (J. Brisou, F. Denis, and P. Babin, 1976). An example of in $vivo$ release of microorganisms by destruction of the substrate occurs when the germs adherent to the gels cross this rampart to penetrate inside the cells, disposing of specific means. For example, the cholerigenic vibrions benefit from neuraminidase identical to the sialidase that attacks the varied glucosidic bonds joining the nonreducer or O-acyl neuraminic residues to galactose, N-acetyl hexosamine or N-O-acyl neuramyl residues of oligosaccharides, glycoproteins, glycolipidic, or colimic acids. This enumeration confirms the breadth of the potentialities of this important enzyme which is also found in the allantoid membrane of the chicken embryo, in grippal viruses, *Clostridium perfringens,* and other bacteria such as *Streptococcus pneumoniae.*

All these observational facts now contribute to a range of arguments in favor of the thesis that microbial adherence, as a whole, is related to the presence of an enzyme. A large number of microbial adhesins attach themselves to specific receptors of a glucidic nature; however, on this level, competition phenomena come into play, notably when identical antigenic patterns converge in a single site. Obviously, a microorganism cannot settle on a site that is already occupied. Yet, in the world of the infinitely small, antigenic communities are very frequent.

Immunologists have observed numerous reactions of crossed immunity. For example, *V. cholerae* and certain *E. coli* strains attach themselves on the epithelium of the rabbit's small intestine and agglutinate the erythrocytes of several animals. The adhesin $\rightarrow$ receptor systems reveal an extraordinary variety in the living world, but often unexpected similarities as well. If the possibly pathogenic bacteria encounter vegetal debris or other reliefs to their liking in a dish, they immediately adhere to them and follow the intestinal transit without spending time on an epithelium. This is a beneficial event for their host, as the "trapped" bacteria become inoffensive. This perfectly fortuitous event could, in a certain way and with a little imagination, inspire the principle of phytoprophylaxia looking for specific sensors of pathogenic bacteria, thus developing a "phyto-trapping" process (Part I, Chapter 1: Adhesins).

These acquisitions are of interest for those who want to understand certain relations between food and intestinal imbalances. We should recall that the numerous and varied vegetal lectins play an important role in adherence. Some lectins, while attaching or agglomerating bacteria, oppose themselves to their attachment on epithelia. Others, on the contrary, can act in the opposite direction and facilitate adherences. Concanavalin A, for example, inhibits the adherence of lactobacilli. Jones[179] showed that certain lectins prevent the attachment of *E. coli.* The studies by Brady et al.[180] (1978) pertaining to the lectins identified in human fecal matter reveal that wheat lectins specifically retain the patterns containing N-acetylglucosamine. In principle, it will therefore attach any particle or molecule carrier of this chemical pattern. The lectin of *Dolichos biflorus* adheres specifically to the nonreducer terminal residues of I-N-acetyl-D-glucosamine. The agglutinin of *Lotus tetragonolobus* (asparagus seed extract) acts on the nonreducer endings of D-galactose. The *Ricinus communis* lectin unites specifically with the nonreducer endings of galactosyl.

These facts confirm the essential role of the polymers studied in Part I. The reactions between these polymers, the solvents present, and the ionic effects are now well known and evidence the stereochemical and molecular specificity of these phenomena. Another factor dealing with local immunity also plays an important role in these adherence processes to the tissues. This already old concept has been fully confirmed. The antibodies of intestinal mucosa, or of buccal mucosa, for example, are opposed to the attachment of a certain number of bacteria. The local antibodies provoke agglutinations, or even pure and simple destructions of the bacteria. There are, nevertheless, circumstances in which microorganisms such as streptococcus defend themselves alone by modifying their ectostructure and by covering themselves up, for example, with a layer of hyaluronic acid that will enable them to resist the phagocytosis.

The welcome that a mucosa may reserve for a microbial population is subject to eclipses. In the world of the living, where nothing is simple, we are dealing with relations of otherness — a world where the strange and the stranger dominate with situations of acceptance and refusal. The relations between microbes and their hosts require us to not only take into account ectostructures, or interfaces more or less rich in specific patterns, but also a whole group likely to modify the possible host.

A doctor understands the importance of his or her patient's constitution, their overall condition, immune potentialities, or hormonal balance. Like the biologist, he or she is now informed of the role of adhesins and mucous gels whether structural or soluble. Everything comes into play in the processes of adherence, whether these processes succeed or fail. On the other hand, we must not forget that mucosae are protected by an especially substantial native microbism in the digestive tube and the mouth.

Despite the role played by natural factors, some microorganisms manage to attach themselves. Experiments (Freter et al., 1978–79)[181] conducted on the behavior of *V. cholerae* confirm that the nonchemotactic strains do not contract any bonds with the mucogel of the intestinal mucosa whether *in vivo* or *in vitro*. However, the chemotactic mutant strains rapidly colonize the epithelium of axenic mice.

A certain amount of data based on research dealing with *E. coli* isolated from a serious and accidental human infection enable the conclusion that the essential part of the infectious mechanism remains the aptitude of the germs to associate with the intestinal muco-gel and not their attachment, in the strict sense of the term, on the epithelium, at least within the restrained limits of experimentation. This conclusion cannot justify a generalization, but it actually suggests just the opposite, since in certain circumstances this bacteria-gel association is shown to be harmful. It is noteworthy, within the framework of these generalities, to recall the role of microtubules that participate in ionic transfers. Certain authors used the following for models: the vesical epithelium, mammal kidneys, fish branchia, or the small intestine. Mucous cells are endowed with these microtubules and microfilaments, notably on the small intestine. Bader and Monet[32] (1978) experimentally showed *in vivo* that agents, disintegrating microtubules such as Vinca alkaloids, significantly modify the transfers of Ca^{2+}, Na^{1+}, phosphates, and water, which confirms the participation of these microtubular formations in ionic exchanges. It is likely that these alterations also modify the behavior of microorganisms, bacteria, viruses, and yeasts on membranes, and that they are likely to promote or hinder adherence.

Conclusively, in all microbial attachment processes on a mucosa or any cellular surface, the concentration in microbial cells and their abundance in adhesins should be taken into account, while understanding that these appendages are themselves submitted to the imperatives of genetics and likely to mutate. The importance of the ionic composition, and thus the electric charges and the nature of the immediate environment, should be noted. The observer should understand the importance of the number of receptor cells, with elements on their membrane structures likely to temporarily or permanently unite with microbial cells. These are naturally glycocalyx, a variety of lectins, complex glycoproteins, etc., studied in the previous chapters. Despite the abundance and complexity of the reactions between all these factors, adhesion and adherence are nonetheless the dominant elements of microbial life in living organisms, whatever kingdom they belong to. The particular case of erythrocytes is, from this point of view, interesting. In this type of cell, the carbohydrate complexes play a major role. Hemagglutination is a phenomenon regularly observed in adherence checks and the detection of molecules of specific recognition. This is a good opportunity to recall that these red cells are covered, among other things, with protein, glycoprotein, and lipidoglucoprotein complexes. The originality of A and B blood groups is their having a sugar that enters into the structure of a surface glycoprotein. This sugar occupies the end of the glucoconjugate chain.

The specificity of Group A is an acetylglucosamine. Group B owes its specificity to a single galactose. The appropriate hydrolases, such as N-acetyl glucosaminidase for Group A erythrocytes and galactosidohydrolase for Group B, try to eliminate these specific patterns and abolish the originality of the cells that enter straightaway into Group O. This fact has been observed not only in humans but also in many animals. As a result, any modification made in the composition of the interfaces causes very different behaviors in the reactions and relations between bacteria, parasites, viruses, and erythrocytes. Lipoteichoic and teichoic acids, as well as lipopolysaccharides, also intervene in all these mechanisms and are part of the cellular receptors. Their position facilitates encounters.

Mucins, which are widespread in living beings, make excellent captors of microorganisms as mentioned previously. They participate, for example, in the structure of the peribuccal mucus of mollusks and notably retain enteroviruses and some bacteria.

2 ADHERENCE TO A FEW ORGANS AND ORGANISMS

1. MAN AND ANIMALS

The limited framework of this monograph only allows for the presentation of a few concrete cases, confirming *in vivo* the consequences of adherence to tissues. These aspects concern human as well as veterinary medicine. For approximately 20 years now, the relations between attachment of bacteria and pathology have been the subject of colloquiums and publications worldwide. The mechanisms of this attachment are increasingly known and lead to practical applications including the preparation of cellular coat-based vaccinations, such as fimbriae, capsules and other adhesins. C. Forestier et al.[182] studied the adherence of an enteropathogenic *Escherichia coli* strain isolated in Burundi (Africa) and analyzed its behavior on two types of culture cells. Only 6 strains out of the 69 experimented on adhered to the brush border of human enterocytes. The authors suggested that these *E. coli,* although responsible for severe gastroenteritis, probably have different, still unknown systems of adherence.

In any case, it is difficult to assimilate what is observed *in vitro* on tissue cultures to what happens in nature on a whole intestine and wild strains. Many authors refer to the intervention of genetics on which the biosynthesis of the attachment elements depends. Birkhead et al.[183] (1979) monitored the behavior of buccal *Neisseria* and the biosynthesis of their polysaccharides. An alpha glucan with a predominance of bonds (1–4) and bonds (1–3) appears with ramifications (1–6). The basic material is composed of two polysaccharides, one close to amylo-pectin represents the main one; the other is an alpha (1–3) glucose. These products are synthesized as always under genetic control.

1.1. Microbisms of the Mouth and Teeth

Buccal microbial populations vary from one individual to the next, depending on age, lifestyle, diet, and hygiene. These microbisms are substantial in number and variety. Bacteria take an active part in the formation of dental deposits. Some possibly pathogenic germs concentrate on the gums, multiply, ferment, and are at the origin of pathologies ranging from basic caries to gingivitis and pyorrheas, Fauchard's disease, and other stubborn pain. On this level, adherence is directed by a series of enzymes, notably sialidases, which are better known by the name of neuraminidases and are responsible for the reaction:

$$\text{sialoglucoside} + H_2O = \text{sialic acid} + \text{glucoside}$$

The insolubility of this glucoside makes it partially responsible for the build-up on dental enamel. Very widespread in the bacterial world, this enzyme is found in *Streptococcus pneumoniae*, *Streptococcus mitis*, *Clostridium perfringens*, *Vibrio cholerae*, lactobacilli, *Myxovirus influenzae*, etc.

Other enzymes belonging to the group of dextranases, levanases, glucosyltransferases, peptide synthetases, and cycloglycases intervene in this process of bacterial adherence on the teeth and gums. The metabolism of saccharose (Part I, Chapter 2) leads to the construction of dextrans and levans. *Streptococcus mutans* and *Streptococcus sanguis,* actinomycetes act thanks to these activities. The accumulation of cellular debris, microbes, and food in the area of the dental and gum palisades creates ideal conditions of welcome for all the microorganisms penetrating in the buccal cavity — gingivitis, *pyorrehea alveolaris*, arthritis, and caries are the direct consequences. (See Figure 35.)

Pliny, the Elder, wrote nearly 2,000 years ago (Natural History, Book XXXI-101) that "the teeth do not have caries, or rot, if every morning, before eating, salt is held under the tongue until it dissolves". Modern specialists continue to lavish this same advice! As early as 1887, Calippe had established the unquestionable relationships between the build-up of detritus and microbes on the teeth and the appearance of diseases. The salivary flow bathes this system permanently and brings it all the nutritive elements. There are differences in composition between the sublingual, parotid, and submaxillary salivas. One of the essential properties of these salivas hydrodynamically remains its viscosity and its wealth in mucus. They play a protective role but any slowdown of flow, any stagnation on the gingival grooves makes them a genuine bouillon culture medium, favorable to all kinds of microorganisms. The average density of bacterial populations is estimated at 10^5–10^8 cells per ml. Rutter[184] monitored the variations of these values in function with periods of activity and rest. Teichoic acid is recognized as playing an important role in these attachment mechanisms, notably concerning the protein M of streptococci, associated with fimbriae and thus with a structural lectin. As shown in the studies by Swanson et al.[185,186] in 1969 and Beachey and Ofek[187] in 1976, *Streptococcus pyogenes* have a serious stock of adhesins and protection systems for massively colonizing the epithelial cells of the buccopharyngeal region, while residing in optimal zones that were inspected by electron microscopy.

The hemolytic streptococci of group A have a much larger range of activity since they are responsible for generalized affections: septicemia,

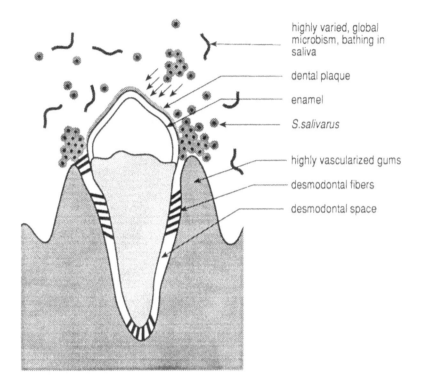

highly varied, global microbism, bathing in saliva

dental plaque

enamel

S.salivarus

highly vascularized gums

desmodontal fibers

desmodontal space

Figure 35.   Bacteria and teeth. Commonplace bacteria elaborate their adhesins and attach themselves to teeth. *S. salivarius* colonizes the gums and secretes its glycanes. It releases fibers and polymerization enzymes. In the organization of a layer, a genuine mesh can be seen. Dental plaque will itself be colonized and become a hiding place for numerous microbes. This build-up will be at the origin of a pathology ranging from the basic cavity to gingivitis, pyorrhea, etc.

scarlatina, erysipelas, abscesses, etc. By using labeled bacteria, Beachey et al.[188] (1977) showed that thrombocytes, erythrocytes, lymphocytes, and desquamous epithelial cells have specific receptors favorable to lipoteichoic acid.

Gingival hemocultures enable another view of this entire parodental pathology. They were suggested in 1930 by R. Vincent;[189,190] the technique is one of the simplest and the results outstanding. The author compared the gingival tissue to a genuine "emonctory", eliminating the toxins and bacteria carried by the blood stream, which is the departure point for many infectious foci. The intestinal origin of certain buccal bacteria was suspected as early as 1927, then confirmed by Antoine[191] in 1939. The research by A. Paoli et al.[192] (published in 1948) also concerned this pathology. It establishes the relations between the alveolar pyorrhea and certain chronic intestinal affections with possible repercussions on the endocardium. (See Figure 36.)

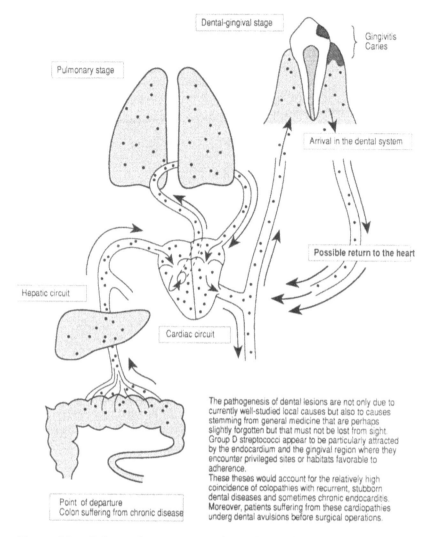

The pathogenesis of dental lesions are not only due to currently well-studied local causes but also to causes stemming from general medicine that are perhaps slightly forgotten but that must not be lost from sight. Group D streptococci appear to be particularly attracted by the endocardium and the gingival region where they encounter privileged sites or habitats favorable to adherence.

These theses would account for the relatively high coincidence of colopathies with recurrent, stubborn dental diseases and sometimes chronic endocarditis. Moreover, patients suffering from these cardiopathies underg dental avulsions before surgical operations.

Figure 36. Schematic representation of relations between chronic endocarditis-colopathies and odontopathies (caries-gingivitis).

These studies lead to the conclusion of an obvious predominance of type D streptococci, *Streptococcus faecalis* and their repercussions. Along with G. Moustardier[193] and other colleagues, we participated in this type of research and performed a certain number of gingival hemocultures in 1951 and 1952. *S. faecalis* were frequently isolated. The microbes circulating in the gingival tissue cause genuine microembolis in the capillaries and lead to either local damage or remote scattering, in particular on the endocardium and the cardiac valvulae. Cardiovascular surgery specialists reasonably prescribe dental extractions for subjects whom they are about to operate on. Postoperative streptococcal septicemia do indeed count among the most fearsome complications.

All in all, the mouth and the pharyngeal region shelter an impressive number of microorganisms. Halves of hydrocarbonated molecules, glycoproteins, take part in wrapping epitheliums and dental surfaces, possibly protecting them from adherence. However, the bacteria strike back by developing patterns in all points similar or even identical to the antigens of the blood groups (Gibbons et al.,[194] 1975). Covered in this way, these bacteria proliferate easily since they have, in a way, integrated into the organism where they settle. Moreover, they can transmit this new identity to other microorganisms, notably the antigenic communities of erythrocytes and mucous cells (Sonju,[195] 1977).

Similar transformations are found in parasitology — Pautrizel's "harlequin" parasites. The importance of extracellular polysaccharides should be noted in the successful attachment of germs on teeth. The field of dietetics seems to have understood; nevertheless, there are some uncertainties. *S. sanguis*, for example, a producer of dextran, adheres on hydroxyapatite covered with saliva, whereas *Streptococcus salivarius* is more likely to attach to a bare substrate.

1.2. Microbisms of the Digestive Tube

Microbisms of the gastrointestinal tract are unquestionably the richest in the number of populations and varieties whatever the organisms concerned in humans or animals. As paradoxal as it may seem, we are still rather unfamiliar with them. The majority of the studies devoted to this organ, which is itself complex, are limited to analyses of fecal matter, overlooking the substantial quantity of populations attached to mucosa. It should be acknowledged that many authors currently perform mucous samplings during fibroscopies, which means unquestionable and instructive progress is being made. With the cooperation of Boisson and Brangier,[196] we published a study (1977) dealing with 27 patients suffering from chronic colopathies. Numerous bacteria were isolated by analyzing fragments of mucosa and mucus sampled using coloscopy. There was a predominance of *Pseudomonas* of rather rare species that certainly escaped the analyses of usual stool.

These studies still remain relatively limited, except perhaps in gastric and duodenal explorations. Each level of the stomach — digestive tract, small intestine, colon, sigmoid, rectum — is characterized by a specific structure selecting a relatively original microbism. Villosities, microvillosities, secretory cells, mucous cells, Paneth's cells, crypts of the small intestine, and enterocytes of Lieberkhün's glands all have a special role to play in these adherence mechanisms. The latter cells migrate from the basal membrane to find the intestinal exit.

Reproduction and migrations take place every 48 hours, and this point will be developed shortly. The villosities are bristled with microvillosities where electron microscopy shows glycocalyx. A genuine mucous film rich in glyco-proteins is often sulphated and synthesized by the mucous cells covering the epithelium. Under the action of peristalsis, this mobile layer heads towards the large intestine, makes it to the colon, the sigmoid ansa, and finally, the rectum. All these segments are covered with a highly wrinkled mucosa, offering numerous welcome sites for an extremely varied microbism where bacteria, viruses, yeasts, and other microorganisms, including parasites, cohabitate. Irrespective of the observational point chosen, there is

- A normal, native intestinal microbism, which is both essential for the system's balance and a real protector of the mucosa.
- A transient, migratory, or momentaneously sedentary microbism.
- An undesirable, pathogenic microbism.

This bacterial crowd, which is composed of thousands of cells, is subject to variations as previously indicated. A rapid review is sufficient for tracing the colonization modes of the mucosa on the various levels.

It is obvious that these techniques, often cumbersome, should not exclude the practice of conventional coprocultures, which have widely proved their worth and remain indispensable. Each observational level has its value, but no single one can claim superiority. The observer needs to know the limits and interpret the results based on these values and research programs. (See Figure 37.)

1.3. Microbisms of the Stomach

The gastric mucosa is willingly inhabited by Gram-positive bacteria, and notably by lactic bacteria. In rodents, the keratinization of a portion of the organ can explain this singularity. As the germs are perpendicularly attached to the mucosa, one of the poles remains free in the gastric lumen; the ambient acidity promotes the growth of these microbes. The adhesins involved in the attachment belong to the group of mucopolysaccharides. Currently, a great deal of interest is focused on a set of very particular bacteria which previously belonged to the *Vibrio* genus (Smith and Taylor,[197] 1919) due to their substantial mobility. In 1963, their curvy appearance caused Sebald and Veron[198] to give them the name of *Campylobacter,* taken from the Greek *campulos,* meaning curve. Lastly, as electron microscopy revealed their helicoidal nature, they became *Helicobacter.* This new genus currently brings together a dozen species of which many are common to humans and animals.

The *pylori* species remains, for the time being, strictly specific to humans and to the very limited antropyloric area. It is considered as a possible agent of certain inflammations and ulcerations, but absolute proof is still lacking. In a 1994 report, F. Fauchère[199] specified that this bacteria settles on apparently altered cellular junctions and that it triggers immune reactions. The conclusions are based on experiments conducted *in vitro* or on gnotobiotic animals. The germ easily crosses

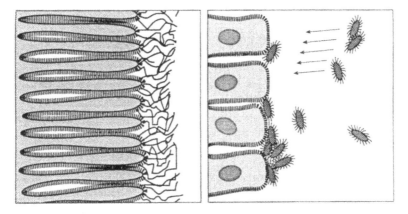

Microvillosities of an intestinal epithelium. Presence of glycocalyx.

Free and attached bacteria forming substantial agglomerates. Union of tissular and bacterial glycocalyx.

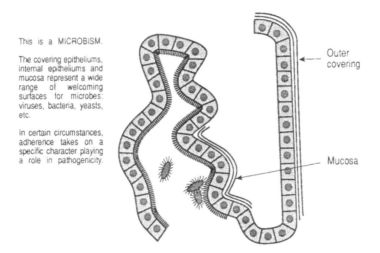

This is a MICROBISM.

The covering epitheliums, internal epitheliums and mucosa represent a wide range of welcoming surfaces for microbes: viruses, bacteria, yeasts, etc.

In certain circumstances, adherence takes on a specific character playing a role in pathogenicity.

Outer covering

Mucosa

Figure 37.   Top — Attachment of bacteria on an epithelium. Bottom — Location on a living organism.

the mucous layer and resists the acidity, due to the activity of a powerful urease that neutralizes the acidity of the antral medium by generating ammonia. The substantial mobility of bacteria enables them to easily reach the cells of the mucosa on which they settle, as shown by electron microscopy photographs.

The antral cells and areas of duodenal metaplasia are practically the only ones to be colonized. *In vitro*, several hemagglutinins have been identified on the cellular receptors as well as N-acetyl neuramyl lactose patterns. An ectoprotein appearing to be associated with urease attaches to a membrane lecithin. Other adhesins unite with carbohydrate sialyl patterns or with gly-colipids. In any case, given current knowledge, it is difficult to explain this specific attraction of *Helicobacter pylori* for human's antral region. An ex-ample of a bacteria adapting specifically to humans — which is certainly not unique — on a particular site is the antropyloric region.

The other *Helicobacter* are much more widely spread out in nature and can be responsible for zoonoses due to their distribution in numerous vertebrates, notably in pets. On the other hand, they are not specialized in a single stage of the digestive tract. The microorganisms continue their trip with the fecal bolus and secretions of all kinds. They reach the small intestine where the living conditions are very different, and there they select other populations.

By observing the process of adherence of a strain of *Campylobacter jejuni* on epithelial cells, A. Belbouri and F. Megraud[200] noted that this adherence is decreased if the bacteria are previously treated with a protease or mannose and L-fructose. However, once again, these are experiments performed *in vitro*. Nevertheless, they confirm the unquestionable action of enzymes that, in the present case, seem to act directly on the bacterial ectostructures.

Filamentous surface species have been studied in depth in rats. The previously mentioned cells of crypts permanently migrate and are eliminated approximately every 20 to 40 hours in rodents. The bacteria that settled on the equally instable substrate solve this "housing" problem by adopting a repro-duction rate faster than that of the carrier cells. Moreover, they emit secondary filaments which enable them to hook onto young replacement cells after leaving the migratory cells. This is an outstanding example of bacteria adap-tation to an instable system. Attachment is ensured via the usual mechanisms previously presented.

The charges of the bacterial surfaces are, in principle, negative (Part I, Chapter 1). The same is true for part of the solid materials present in the medium. The intervention of neutralization factors avoids repulsions of charges of the same sign. The extracellular polymer biosynthesis of anchorage adhesins plays this role. *Pseudomonas*, for example, actively secrete the alginates that enable adherence on any interface. Their frequency in the intestine on almost all levels, and notably on the colic framework, is understood. The receptor glycoproteins of the cellular crypts are clearly less complex than those of the villosities; this fact has been observed in rats and occasionally in humans.

Biochemical activities, which are very different at each level of the diges-tive tube, have an influence on the mechanisms of adhesion and adherence. The relations between the physiological state, the structure of the mucosa, and the synthesis possibilities of the adhesins are not in doubt. By pursuing the gas-trointestinal trip, we arrive in the coecum where purely native and protective spiral bacteria dominate. The fusiform bacteria are, on the other hand, in such abundance in certain animals that they form a genuine continuous film cover-ing the whole mucosa.

The colon and rectum house an extremely rich bacterial population of which one part is also solidly attached to the mucosa and resists the evacuative peristalsis of this area. In animals, these coverings are particularly inhabited by spirochetes. A large amount of research has indeed been conducted in animals or on cultures of animal or human cells, erythrocytes, leucocytes, and mucus. Glycoproteins, glycolipids, and sialic compounds dominate, once again, among those responsible for adherence. What Vosbeck and Helmut Mett[13] summarized in 1981 still remains valid and instructive. The enterotoxigenic strains of *E. coli* and *Pseudomonas* constitute an easy-to-handle, choice material for all these studies. *E. coli* K88, responsible for diarrhea in young porcines, was studied in depth by Jones and Rutter[201] in 1974. The pathogeny was related to an adherence factor named K88 and to the secretion possibility of an Ent+ enterotoxin. These two characters are both controlled by different plasmids. The K88 and Ent+ strains are the only pathogens. They colonize the epithelia and trigger diarrhea. Certain animals endowed with a natural resistance withstand the aggression of these *E. coli* K88 and Ent+. The natural defenses and the absence of synthesis of adhesins combined are both called on for explaining the phenomenon. Some authors even call on a Mendelian-type mechanism. These animals resisting *E. coli* K88 are sensitive to another toxigenic strain 987 (Smith and Huggins,[202] 1978).

The observations made on humans sensitive to certain enterotoxigenic *E. coli* have shown that two adherence factors were also indispensable here for causing disorders. These factors were designated by Evans et al.[203] (1978) by the abbreviations CFA/I and CFA/II (Colonization Factor Antigen). After experimenting on volunteers, it was acknowledged that the strains deprived of factor I are not pathogenic. The fimbriae play an important role in all these mechanisms and contribute considerably to the aggressivity of intestinal germs. Interesting applications have been drawn that will be commented on later. These fimbriae unite with cellular lectins of the host (Levine,[204] 1981). We now know that enteropathogenic bacteria attach rapidly to the mucous epithelia as soon as they arrive in the intestine. In summary, it should be noted that three types of *E. coli* are currently distinguished:

1. Entero-pathogenic toxins that invade the small intestine where they secrete one or more toxins
2. Enterotoxinogenic *E. coli* with a wider spectrum which are found all the way in the colon and are specialized in the production of cytotoxins
3. Invasive *E. coli* penetrating the mucosa which they deteriorate, notably in the small intestine
4. Entero hemorragic

According to Savage,[205,206] *E. coli* is not a producer of lecithinase. However, with my coworkers M. Cl. Bernard and F. Denis, I have studied few strains of

E. coli and shown them to be rich in phospholipase A. The research was performed thanks to bacteria labeled with radioactive phosphorus. (Bernard's thesis — Science Faculty of Poitiers; 1971).[207] Many other bacteria were studied: *Shigella, Chlamydiae*, germs responsible for dysentery in pigs, *Salmonella, Vibrio*, etc. All of them are related to the intestinal mucosa and many penetrate these mucosa where the submucosa secretes toxins when they have the possibility of penetrating into the tissues or stagnating against the mucosa. The mobility of these bacteria facilitates the colonization on gels and the union with the cellular glycocalyx. Native bacteria line all the mucosa and blend with the chemical and physical environment and the production of enzymes for ensuring the proper functioning of the whole. Practical applications have been drawn based on patient observations.

For ages, people have understood the value of certain eating habits, acidifiers, and the regular ingestion of lactic ferments. Commonplace examples are still in the news and are exploited more than ever. Native bacteria specific to the animal species and gastrointestinal sites are very sensitive to medicamentous, chemical, and physical perturbations that can be imposed upon them. These modifications of the local microbism can lead to serious disorders to which we drew attention in 1971 and 1955[208] by describing the sometimes severe consequences of selections provoked by these practices. The antibiotics in use at the time eliminated part of the native microbism, which was recognized as a guarantee of balances and a producer of vitamin and growth factors. This eradication promoted the growth of a sometimes very selected substitute microbism, limited to one or two species which were resistant to the antibiotics and the vitamin deficiencies, justifying the concept of "substitute and selected microbism". Gouet[209] confirmed these observations in 1981, considering that certain antibiotics cause the disappearance of these native bacteria which were essential for the system's balance.

1.4.   Microbisms of the Urogenital Apparatus

The cells of the urogenital apparatus, like those of other organs, are equipped with receptors rich in polysaccharides and glycoproteins. Numerous experiments have been conducted on isolated cells and culture germs. There is often a similarity between bacteria of enteritic origin and those responsible for urinary infections — kidneys included, naturally. *E. coli, S. faecalis,* and *Pseudomonas aeruginosa* are prominent there and, more rarely, staphylococci.

The adherence means of uropathogenic *E. coli* are increasingly well known. The mannose-sensitive fimbriae are found in a certain number of strains responsible for infections. Others fix directly on the glycolipids entering into the structure of the epithelial cells. These potential adherences can be inhibited by specific antibodies present on the mucosa. This concept confirms the value of antigenic therapy in the treatment of persistent infections. These adherences are, as usual, ensured by pili I and therefore fimbriae-lectins, as specified by C. Svanborg-Eden et al.,[210] in 1981. Cell receptors, also compared to lectins, greet them on the surface. (Eshdat et al.,[211] 1978). The relation between the pathogenic power of the bacteria and hemagglutination appears to be due to the recognition by the microorganisms of glycolipids covering the surface of the erythrocytes and uroepithelial cells. These glycolipids belong to the group of surface globotetraosylceramids. These complexes, sometimes made up of long chains, appear to play an important role in the process (Leffler and Svanborg Eden,[212] 1980).

The pili or fimbriae of the uropathogenic bacteria have the same hemagglutinating aptitudes. Fimbriae labeled with radioactive iodine adhere specifically to the cells of the urinary epithelia. Nevertheless, there are sometimes other mechanisms since nonhemagglutinating, mannose-resistant strains also attach themselves to these epithelia. The role of genetics in these adherence capacities has been confirmed several times, in particular by V. Quinn et al. in 1988 on a strain recently isolated from a patient's urine. The bacteria adhered to human erythrocytes and to the epithelial cells in the presence of mannose. The authors managed to isolate the DNA responsible for this property and to transmit it through the plasmid to a primitively nonadherent strain.

The studies made over the past ten years confirm this attachment of the uropathogenic strains achieved through bonds between the bacterial adhesins and the cellular receptors, which are themselves quite varied. P. Courcoux et al.[213] (1988), who studied the behavior of 180 strains of uropathogenic *E. coli* isolated from pyelonephrites, cystites, and that of enterotoxigenic and enteropathogenic strains, showed that three operons, Pap, Afa, and SfA/Fic, intervene effectively in the attachment power of these germs. The genetic information appears to be the condition of adherence of these bacteria to uroepithelial cells — an attachment that seems easier when the cells are altered (Schwartz et al.,[214] 1982). This observation is of considerable practical interest, as it confirms the uncertainties of the values obtained using bacterial enumeration tests in infected urines, loaded with desquamated cells, leucocytes, and erythrocytes. All these cells are covered with germs which form substantial agglomerates. (See Figure 38.)

It is obvious that, in such conditions, the colonies obtained on the usual culture media never correspond to 1 germ but most often to 100, which expresses the amplitude of the uncertainties and the modesty of their value. It is preferable to devote more time to the in-depth study of germs, or the germ responsible for the disease, to recognize their antigenic identity related to the adhesins, as well as the toxicity and sensitivity to antibacterials.

Neisseriae gonorrhoeae offered, in the same way, the opportunity to conduct similar research. The pili were identified on a large number of strains and recognized as the guarantees of adherence, while being submitted to the rules of genetics. *Pseudomonas, Chlamydiae,* and *S. faecalis* are frequently responsible for these persistent infections. It can be noted, since the fact supports our own ideas, that the aptitude of certain bacteria to adherence is

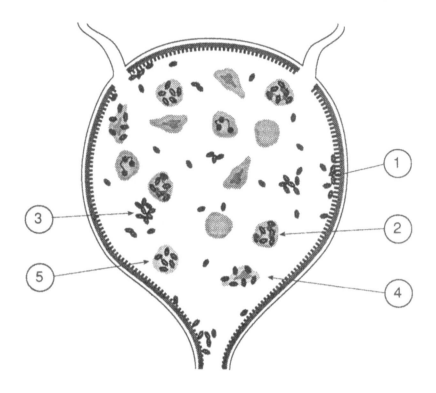

1 - Bacteria attached to the epithelium

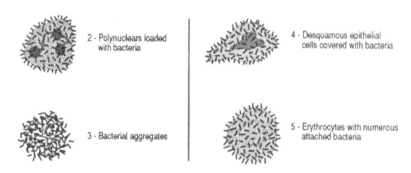

2 - Polynuclears loaded with bacteria

3 - Bacterial aggregates

4 - Desquamous epithelial cells covered with bacteria

5 - Erythrocytes with numerous attached bacteria

Figure 38. Varied position and dispersion of bacteria in an infected bladder, a veritable challenge to enumerations.

decreased if they are previously treated by protease (A. Belbouri and F. Megraud,[200] 1988).

Some enteropathogenic *E. coli* strains have bonding filaments that were identified on the fimbriae and fibrillae by Giron et al.[215] (1991, 1993). On these adhesion filaments, three fimbrial subunits were distinguished. The N-terminal of an amino acid sequence proves to be homologous to that observed on the fimbriae of mannose-resistant uropathogenic strains. Their activity is not inhibited by the alpha-D-Gal (1–4) disaccharide. The authors suggest the existence of relations between the pathogenic power and the structures present on the fimbriae and fibrillae.

This research confirms the frequent intestinal origin of the uropathogenic strains. In order to identify such structures rapidly, in 1991 E. Westerland[216] suggested the use of fluorescent microparticles covered with the adhesin specific to a uropathogenic strain 075x. Experimented on sections of human renal tissue, these particles covered with the adhesin have the same behavior as the bacteria and constitute a loyal reagent enabling a rapid check and the study of adherence.

Chlamydiæ are mandatory parasites concerning the urogenital tract and but also other organs: lungs, eyes, etc. They attach to cells having N-acetyl glucosamine in their surface structure. The initial site of attachment appears to be a protein. A prior treatment of the cells with trypsin preserves them from chlamydial attack (Byrne,[217] 1976). The wheat lectin specific to N-acetyl glucosamine inhibits the adherence of *Chlamydia trachomatis* (Levy,[218] 1979).

Continuation of this inventory is not necessarily worthwhile as it would lead to a treaty of medicine. Irrespective of the apparatus studied, whether ocular, rhinopharyngeal, cutaneous, etc., the same scenario can be seen everywhere: adhesins, cellular receptors, recognition mechanisms between these various tissular and bacterial elements, and specific affinities, examples of which have just been given.

3 NATURAL MICROBIOCENOSES

1. VEGETALS

The wealth of microorganisms in vegetals has been established for a long time. Leaves, stems, and roots of all kinds of plants shelter large populations or cohabitate in animals, bacteria, viruses, lower fungi, yeasts, and microparasites. With C. Tysset, R. Moreau, and C. Durand,[219] we published an inventory of these bacterial communities attached to ornamental plants. Some of these bacteria proved to be experimentally pathogenic for poikilothermic and a few homeothermal animals. The surface of leaves or "phylloplane" is rich in a variety of germs as confirmed by Preece et al. in 1971 and Dicton and Preece in 1978. Humidity and the buildup of organic substances promote the growth and installation of bacterial communities, notably in tropical or temperate regions or in apartments. The dispersion of pollen facilitates remote colonizations.

The external microbism of vegetals was qualified as "epiphytic" (Last, 1955). It is regulated by the environment: humidity, light, temperature, chemical composition, etc. The edaphic conditions are inescapably modified by the new activity that the growing vegetal world creates (Boullard and Moreau).[220]

The phylloplane, covered with various materials, offers conditions favorable to bacterial colonization. At the time of pollen dissemination, vegetal particles constitute both excellent vectors and germ reservoirs. Loaded with microbes, these pollens often land on water, causing massive, natural pollution capable of harming the fauna by eutrophyzation. These contaminants, abundant in putrid bacteria with a high potential of biodegradability, cause, for example, high mortality in fish. Pollen and vegetal seeds promote the spread of diseases. Bacteria attach to these particles, which are often rich in lectins whose role we are familiar with. According to Goodman[221] (1976) and Sing and Schroth[222] (1977), nonpathogenic bacteria, after having penetrated into the leaves, adhere to the cell walls. They are rapidly surrounded by small vesicles and then are imprisoned in a network of fibers. Strains of nonvirulent *Pseudomonas solanacearum* thus infiltrate tobacco leaves. They do not attach themselves to cell walls in the strictest sense of the term since they are prisoners of the

105

gangue secreted by the plant; however, a virulent strain of the same species of *Pseudomonas* rapidly multiplies in the intercellular liquid. Goodman et al.[221] (1977) observed that tobacco leaf extracts infiltrated with avirulent strains of *Pseudomonas pisi* agglutinate the bacteria whereas the experiment performed on a virulent strain does not cause any agglutination.

During similar experiments performed with *Ps. solanacearum*, it was confirmed that the vegetal lectins agglutinate specifically and exclusively the avirulent strains for potatoes and tobacco. The nonagglutinated virulent strains sensitive to the lectins have specific extracellular polysaccharides. This particular role of the lectins has been the subject of several research programs. In 1993, A. Vallace and M. Perombelon[223] observed the behavior of *Erwinia carotovora* on the tissues of potatoes and folios. The alpha-methyl-mannoside inhibitor of the mannose-sensitive strains reduces the attachment on the slices of the vegetable and on the folios. The inhibitor asialoglucide of the mannose-resistant strains is opposed to the attachment only on the folios. The adherences are located on the cellular junctions.

Agrobacterium, an agent of the Crown gall (Part I, Chapter 2), is positioned in an initial stage on a specific site located on the cellular membrane. Lippincott et al.[224] (1969) and Beierbeck (1973) established that the microbial bodies of avirulent strains or phenolic extracts obtained on the spot, containing attachment polysaccharides of the virulent or avirulent strains, were opposed to the development of tumors. It would appear that competitive mechanisms play a role in the appearance of outgrowths. The number of bacteria is taken into account. The O/R antigen of the polysaccharide extracted from strains capable of attaching themselves and leaf preparations can inhibit the tumoral process normally generated by virulent strains. In high concentrations, polygalacturonic acid, pectin, and arabinogalactan also prove to be inhibitors. These basic observations attributed to Lippincott (1978) could account for the infecting specificity of *Agrobacterium* for dicotyledons and gymnosperms and its harmlessness with respect to monocotyledons. The abundance of pectins in the resistant tissues also deserves attention. It is apparently accepted that the aggressive power of the bacteria is a result of the union with the polysaccharides of the vegetal cells. During the different stages of the disease, lectins also play a role.

2. THE RHIZOSPHERE

This term was suggested in 1904 by Hilton for designating the land area directly influenced by vegetal roots. There is naturally reciprocity between roots and microbiocenoses. They cause profound modifications in the structure of soils and their grain size. They develop mineral and organic elements, including lectins and enzymes. Depending on their own metabolism, microbes participate in the life of this characteristic environment of the rhizosphere. Using the buried slide technique, Boullard and R. Moreau[220] (1962) were able

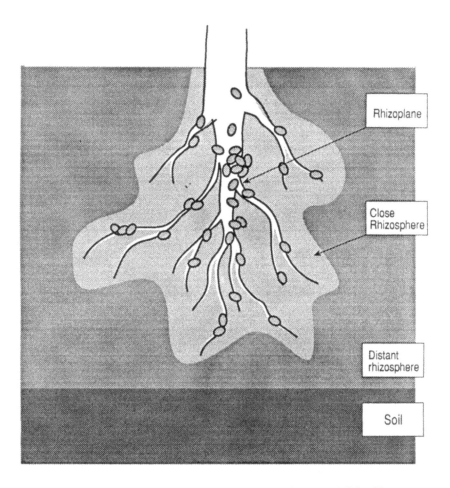

Rhizoplane

Close Rhizosphere

Distant rhizosphere

Soil

Figure 39. **Schematic representation of the relations established between the microbiocenoses of the soil and roots.**

to establish significant counts of this microbism and confirm the variability of it. They evidenced the influence of the roots, some belonging to healthy plants, some to diseased vegetals, on the total microbism. Harley (mentioned by Boullard and Moreau[220]) makes three distinctions:

1. The rhizoplan, corresponding to the roots themselves
2. The rhizosphere that surrounds the roots, and undergoes their direct influence
3. The soil (See Figure 39.)

In 1963, Jenny and Grossenbacher isolated a mucilaginous substance surrounding the roots and spreading into the rhizosphere. "Mucigel" was in certain cases digested by bacteria. In other circumstances, its buildup becomes substantial. This mucigel, rich in polysaccharides, coats both the bacteria and the seeds of the soil, thus constituting a new and very unique biotope where Gram-negative species and Actinomycetals dominate.

Dazzo[226-228] monitored the attachment of bacteria particularly well on this level. The technique of buried slides (Starkey) indicates that in an arid soil, a few bacteria adhere to the easy-to-color slides. It is nevertheless a rather poor and monotonous microbism. However, the slides buried in the area surrounding the roots are rapidly colonized by a variety of microorganisms — bacteria, yeast, lower fungi, protists — that take advantage of this favorable environment. Yet, here again, everything depends on the nature of the soils, the climatic conditions, and especially, the plants themselves. This microenvironment is approximately 0.5 to 1 cm in diameter, taking the roots for the center. The colonization of radicles varies with the segments examined. The microbial communities seem to have a preference for median and proximal zones. Certain bacteria appear to penetrate into the tissues to attach themselves there. *Pseudomonas tolaasii* settle due to the participation, which appears mandatory, of the bivalent ions Ca, Mg, and Sr. The monovalent cations remain without any effect on the progress of the process.

The primordial role of the colonization of the radicles of legumes through the *Rhizobium* constitutes a "particular state of the rhizosphere effect", according to Boullard and Moreau. Noël Bernard and Nobécourt observed that the tubercles of orchids secrete a fungicide substance limiting the extension of their normal fungal associates.

Boussingault, and then Ville (1850) should be given credit for the concept of atmospheric nitrogen attachment through nodosities which they observed on the roots of legumes. Later, with the progress in microbiology, it was recognized that these nodosities were generated by symbiotic bacteria which were identified by Frank in 1889. He named them *Rhizobium*. There are capsulated forms of these Gram-negative germs located in the family of *Rhizobiaceae* and covered with fimbriae. They all have the property of causing the formation of nodes on the roots of legumes — beans, lentils, lupins, clover, etc.

This production results from a series of events intimately related to the conditions of the environment. Certain hosts can be indifferent, or in a state of "tolerance", whereas others accept the symbiotic life without any problem. Certain nodes which are normal in appearance nevertheless remain functionally inactive. Lastly, the *Rhizobium* can go from the symbiotic to the pathogenic state which some have compared to an "alliance reversal".

To understand these events, we need to re-examine the importance of the structures of the energetic elements present in the environment, the grain size, and numerous factors, on one hand resulting from the bacteria themselves, the variability of which is known, and, on the other hand, the cellular receptors. In 1975, Fahraeus[225] noted a deterioration of the cell walls on the radicles after an attack of pectinolytic enzymes of bacterial origin. This destruction begins at the attachment point of the *Rhizobium*. (See Figure 40.)

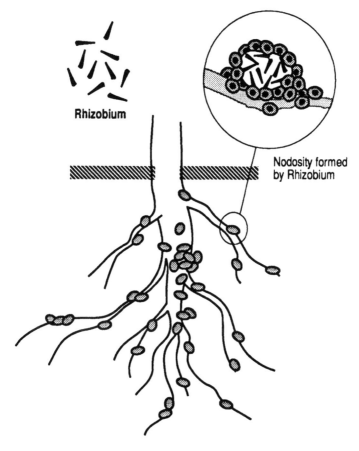

Rhizobium

Nodosity formed by Rhizobium

Nodosity on a leguminous root

Figure 40.   Population of leguminous roots by Rhizobium.

In 1976, Dazzo[226-228] managed to quantify the phenomenon and determine the specific character of the attachment, which was a function of the bacterial strains and plants examined. Molecular phenomena govern the whole with a recognition stage of the germ by a lectin. The N-acetyl galactosamine inhibits adherence, and its relations with hemagglutination were also pointed out.

We are tempted to conclude here, as elsewhere, that the specificity and originality of many events offered by life, irrespective of the kingdom, family, species, or genus, are found in molecules and stereochemistry. The glucidic patterns are practically always present. Nevertheless, the bonds between lectins and glucidic patterns do not account for the infectious process all by themselves. It would be overly restrictive to attribute everything to one cause. Some authors, for example, consider that lectins act as basic chemical mediators. It should be noted that they are also "bacteria receptors". The definitive adherence and especially the invasion of the plant by bacteria are a result of more complex mechanisms. Relations exist between the roots, diatoms, fungi, and yeast more or less associated with bacteria. These interactions cannot be neglected. The *Rhizobium* occupies a privileged place in the attachment mechanisms of atmospheric nitrogen, but the possible role of other microorganisms cannot be excluded. Endotrophic and ectotrophic microbes participate in the process at other levels.

In sum, it is recognized that lectins, which are abundant in seeds but are also present in other parts of plants, often play a role in the progress of all these processes. There is no doubt about their protective action, but their attitudes are sometimes of disarming contradiction. Their quasi-universal presence leads to a rapid analysis of these sometimes surprising behaviors. They are summarized in Figure 41, accompanied by figures that will put an end to these considerations, the practical interest of which cannot go unnoticed by anyone undertaking an in-depth analysis of bacterial life in a soil where releases, guided by fundamental data, are useful. Examples of them will be given in the technical part.

3. SOILS AND SEDIMENTS

The plural of soil and sediment should be used straightaway due to their extreme diversity. Soils and sediments live thanks to the microbial populations that they have sheltered for thousands of years. Inert, sterile soils are extremely rare. Microorganisms live as frequently in polar regions or near volcanoes as in extreme environments of all kinds. To be convinced of the fact, it is sufficient to use global enzymatic methods or oxidoreduction reagents such as teatrazolium salts or bioluminescence. These basic techniques testify to a vital activity in the soils and sediments that breathe and ferment on a permanent basis thanks to the microbiocenoses they shelter.

Practically speaking, the microbiology of soils was born around 1885 with Beijerinck,[117] Winogradsky,[21] and their colleagues. Microorganisms are obviously attached to the soils — "life depositories" — according to J.C. Gilman (1957). The heterogenous character of these environments explains the variety of the populations. Composed of inorganic matter ranging from massive rocks to the finest clayey sediments (Montmorillonites; kaolins, for example), they are also rich in organic matter such as animal and vegetal debris, which in certain areas represents over 50% of the whole. The microbism living on this heterogenous whole leads it to the humidification that will play an essential role, of which we should never lose sight, in vegetal life as well as in adherence.

The polyphenol-based humic fraction, rich in carboxyl groups and fulvic and humic acids, results from the hydrolysis of lignin. These substances constitute large food reserves for bacteria. Although humic acid is insoluble in an acid medium, fulvic acid can be dissolved as easily in an acid medium as in an alkaline medium. In the attachment mechanisms of bacteria in the soils,

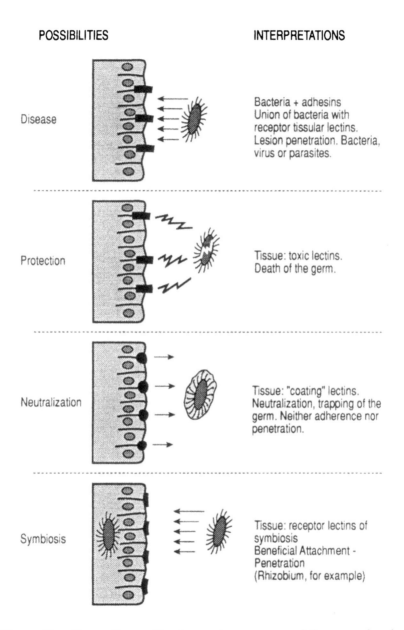

POSSIBILITIES   INTERPRETATIONS

Disease — Bacteria + adhesins. Union of bacteria with receptor tissular lectins. Lesion penetration. Bacteria, virus or parasites.

Protection — Tissue: toxic lectins. Death of the germ.

Neutralization — Tissue: "coating" lectins. Neutralization, trapping of the germ. Neither adherence nor penetration.

Symbiosis — Tissue: receptor lectins of symbiosis. Beneficial Attachment - Penetration (Rhizobium, for example)

Figure 41. Depending on the circumstances, one and the same chemical and molecular affinity can be represented by a protection or the opening of a door to an unwanted bacteria.

the ionic phenomena play an important role. The pH values and the dissociation constants (pK) of the membrane components are taken into consideration. These aspects were studied on a few bacteria (*Klebsiella, Staphylococcus aureus, Streptococcus pyogenes*) at different pH levels and in function with their electrophoretic mobility (Jones, 1972).[229]

The interest of teichoic and lipoteichoic acids is recognized, entering in the composition of the cell walls. Marshall[230] (1967) monitored the behavior of *Rhizobium* with similar methods. At a high level of acidity, the bacteria becomes cationic. The surface carboxyl groups in *Rhizobium lupini*, dissociated with pH 4, no longer exist when this pH level goes under 2. The cellular surface loses its charge. Depending on the value of the pH, the bacterium, due to the dissociations of the surface components, changes signs. For example, if it is positive at pH 2, it will be negative above pH 4.

It is obvious that the attachment mechanisms of the bacteria in the soils are intimately linked to the structure of the host substrate where inorganic elements and organic matter mix together without any apparent order. Once the ionic adsorption is completed, the succession of the following stages, as far as bacteria is concerned, will follow the routes analyzed in the previous chapters. After having abided by the laws of physics, in the heart of the humus the bacteria will adhere and incorporate themselves into substrates and live there.

These same processes are found in biofilms. As specified by R. Burns[231] in 1979, "it is difficult to imagine a more complex environment than a soil, where apparently infinite chemical, physical, and biological forces interfere on a permanent basis." The bacteria act on the molecular and colloidal level as well as in the form of aggregates. "The environment selects the microbes and the microbes create the environment" (R. Burns). The result is that the microbiology of soils can only be understood by constantly taking into account the structures of the terrain, its composition, the ambient physical conditions, meteorology, etc. All these entities transmit their specific qualities, their scent, and their taste to certain plants — vines, to name one of many.

It is of utmost importance to insist on the fact that in soils, bacteria are practically always organized in communities and aggregates, to which Winogradsky[21] devoted an especially attentive analysis (and rightfully so) by expressing the doubt about the value of enumerations. That is the reason why he focused his research on the overall enzymology of dirts. His methods have been widely developed since then by all his disciples, with all the microbiologists devoting interest to soils and sediments. We used these methods extensively in 1982 with R. Moreau and others.[232]

Microorganisms divide up the chores. Schematically, two sorts can be distinguished:

1. A "zymogenic" microbism, acting rapidly and immediately to "digest" any massive and abundant arrival of unbalancing substrates.

2. A more complex, native microbism which develops in a second stage, with the association of bacteria, yeast, lower fungi, and even protists that will act more slowly to complete degradations and fertilize the soils for the good of the vegetals.

A.G. Lochhead[233] took into consideration the food requirements of attached bacteria. It is thus easier to understand these successions of populations on a given biotope and to confirm the close relations established between these populations and the environment they occupy. B. Boullard and R. Moreau[220] (1962) applied these principles to study metabolic cycles on plants and soils, notably forest soils.

In the end it should be retained that soils, even in extreme environments as far as the biosphere is concerned, have been the main source of microorganisms ever since the appearance of life on earth. They are found in aquatic environments where they participate in the vital cycles and in the metabolisms of the organic matter with perhaps more variety. During these processes of degradations and reformations, where nothing is lost, adherence is a necessary condition. It provides the contact and activity of the microorganisms on their targets.

4. AQUATIC ENVIRONMENTS

Oceans alone occupy 70.8% of the surface of this planet. If lakes, rivers, and glaciers are added, as well as what they contain in inert materials, it can be concluded that this entity constitutes the largest source of microorganisms in the biosphere. This chapter will be devoted to bacterial microbiocenoses which, installed on all interfaces form numerous biofilms that are slightly different from those observed on living tissues. Wherever microorganisms are attached, whether on living tissues or on inert surfaces, microbial deposits construct genuine carpets known as biofilms. Living beings such as plankton, invertebrates (notably mollusks), and aquatic plants, but also extremely varied inert materials that shelter vast bacterial populations, can be found. Industrial materials are included in this group.

Roth[5] lent particular interest to underwater surfaces. A polysaccharide film rapidly invades these surfaces, the bacterial origin of which is not questioned in the least. Electron microscopy shows that microbes are solidly anchored there due to a fine network of polysaccharide fibers. Inside this network, an increasingly complex colonization can be seen. It was estimated that a surface submerged in a stream of water can attach approximately 10^6 germs/cm^2, whereas water only contains 10^2 to 10^3/ml. The difference is therefore significant.

Martial (1978) insisted on the importance of interfaces, present everywhere in nature, confirming the ancient concept that the majority of microorganisms, aquatic beings, require an aqueous environment to develop. Any discontinuity in a system creates a "surface", or better still, an interface, capable of modifying the behavior of bacterial populations, even microbiocenoses taken as a whole. These interfaces separate: solids/liquids, solids/gases, liquids/gases (air, for example) and non-miscible liquids: water/hydrocarbons, water/oils, etc. (See Figure 42.)

Each of the phases taken separately has particular chemical and physical properties. Any surface, whatever the nature, plays a crucial role in the microbial life of an ecosystem. It represents a pole of attraction and a potential source of nutritive elements, and it favors the deposit of the most varied organic or inorganic materials present in the environment.

Germs colonize the submerged surfaces very rapidly. Roth[5] experimentally confirmed these already old concepts. Surfaces submerged in salt water or fresh water have been the topic of numerous experiments. Grains of sand, metal sheets, glass plates, and plastics have been examined hundreds of times using either an optical or electron microscope. We lent interest to the colonization of metals by microorganisms for a number of years with several colleagues (J. Brisou, 1967–68, 1972–73)[234-236] and thus monitored the bacterial corrosion of these materials. Colonization is always rapid and substantial: macromolecules absorbed in an initial stage modify the physical properties of the surface. This first layer promotes the installation of microorganisms. High energy surfaces thus facilitate the formation of a possibly welcoming film. Gravel, residue of all kinds, fine sand, and suspended matter participate in the process. In water that is poor in organic matter, the quantity of free bacteria remains relatively low. They are, in general, attached to the media having concentrated a few energetic elements present in the microenvironment through absorption.

At the end of the last century, Duclaux[1] had already written that microbes form a film on the surface of each particle and participate in the biological balance of the ecosystem. However, here everything depends on the structure of the media — their relations as shown by Flechter[237] in 1979 and Burns the same year. Montmorillonite, often given as an example, represents one of the effective clays. The average diameter of the particles is around $2\ \mu$. There appear to be 90×10^9 for 100 mg, providing an absorbent surface of 11×10^3 per cm and per B.

Clays therefore constitute imposing absorbent surfaces enabling the concentration of organic and mineral nutritive substances particularly favorable to microbial colonizations of all kinds. We were able to determine 4.5 to 9 mg for 100 g of total organic matter on medical-use clays, kaolins, attapulgite, etc., in principle purified since designed for treatments. These materials retain water and enhance certain hydrolyses. They retain toxins. Sometimes antibiotics capture cations, microorganisms, and viruses (J. Nestor, J. Brisou, F. Denis et al., 1964, 1975, 1986).[18,19] Coxsackie viruses are thus absorbed at 100% by montmorillonite. The captive role of grains of sand was particularly well analyzed by Meadows and Anderson[238,239] in 1968, and by Z.

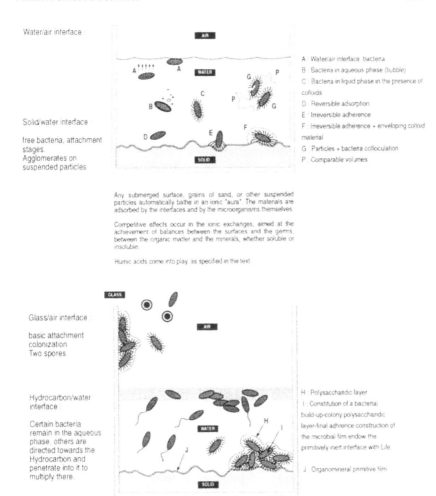

Figure 42. **Behavior of bacteria in the presence of different types of interfaces.**

Dartevelle-Moureau[240] in 1976. These authors monitored the evolution of microorganisms on sand from beaches and showed the massive colonization of these microsurfaces. (See Figures 43–44.)

The images reproduced here (See Plates 1, 2, and 3 and Photo 1) give a concrete example of these situations. Alexander[2] confirmed this massive colonization by bacteria, yeast, lower fungi, viruses, and protists of all the submerged, solid interfaces. Emphasis should be placed on the fact that the analysis of sand from beaches is much more useful and instructive than the basic water analyses as far as the sanitary supervision of vacation areas are concerned. We have described the techniques in detail and recall that the first attempts of enzymatic release were performed precisely on coastal sand in

The location of microorganisms in nature is related to the interfaces:

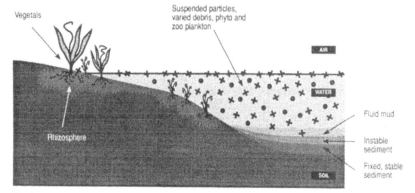

The free microbes, in the strictest sense of the term, in this ecosystem constitute the minority (approximately 1/100, 1/1000).

The remainder of the microbiocenoses, for vital reasons, populate all the inert and living interfaces. They exert their maximal activity there, find shelter, food and favorable conditions for reproduction.

This concept is widely consolidated by all the current surveys and data of microbial ecology.

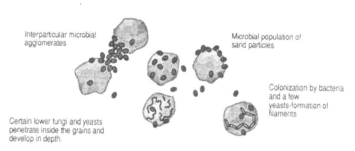

According to Meadows and Anderson (1968), 5 to 150 cells are counted per microbial colony attached on the surface of a grain of sand.

Figure 43.   Top — Aquatic ecosystem. Bottom — Colonization of grains of sand. Although some of these colonies correspond to a single apparent species, there are nevertheless heterogenous bacteria, cyanophyceae, and yeasts.

1978 and 1979.[241-242] By using oxidoreduction reactions, it is also easy to control the activity of sedimentary or attached microbiocenoses on plankton.

Sedimentary water — or plasma — is extractible by ultra-centrifugation or forced filtration. It constitutes a solution of organic and mineral substances facilitating microbial life. The type of sediments obviously varies depending on their location, which makes it possible to differentiate stable coastal sediments, instable sediments, constantly submerged sediments, and estuary sediments

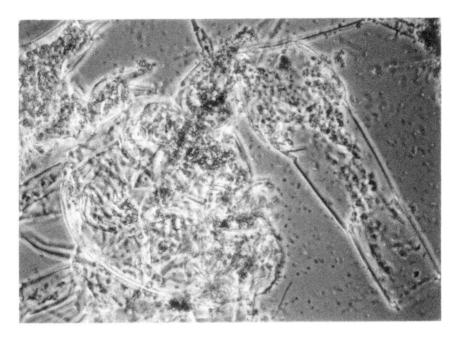

Plate 1

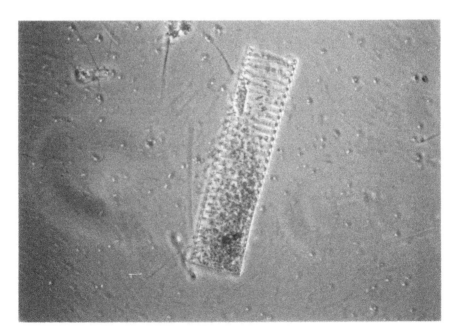

Plate 2

Plates 1, 2 and 3.   Bacterial populations on marine plankton. Substantial oxidoreduction activity demonstrated by tetrazolium-red formazan reaction.

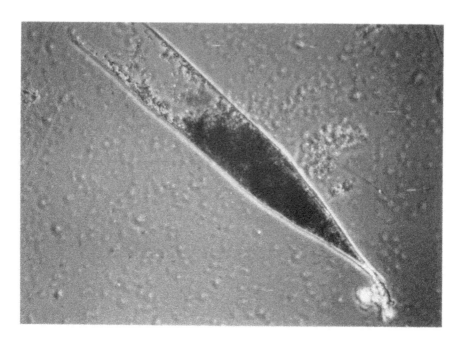

Plate 3

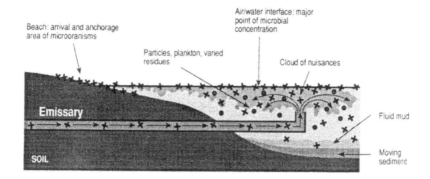

Figure 44. **Arrival of an emissary in an aquatic ecosystem. The cloud of nuisances rises quite naturally and rapidly towards the surface where, obeying the conventional laws, the particles of all kinds spread out, carrying out ascending and descending movements. The aquatic medium shelters numerous very active microorganisms, phyto, and zoo plankton, of which the mobility, horizontal and vertical travels are known for vegetals, invertebrates, and vertebrates. All of them, without any exception, condition the evolution of the chemical and microbial pollutants unloaded into the media. The intense life launches a permanent challenge to all the mathematical forecasts. A simple sunbeam, or on the contrary, a cloud or rain, are enough to modify the behavior of the zoo plankton, the main vector and captor of microorganisms. Such events totally change the breakdown of microbiocenoses.**

(Dartevellee-Moureau; 1975)[240] such as very mobile and instable fluid muds and turbidity maxima. All these models differ from deep sediments of the hadal or abyssal zones. Each one, depending on its location and structure, shelters more or less specific microbiocenoses. Alongside specialized barophilic and psychrophilic germs, observers encounter commonplace, ubiquitous yeasts or bacteria, even at very substantial depths (ZoBell,[243] Kriss et al.,[244] J. Brisou et al.[245]) All these microorganisms are, in principle, attached to the microenvironments analyzed here. The adherent bacteria most often lose their usual morphology in culture media.

It should be understood, once again, that the life of germs in the wild differs from that observed in the laboratory and can be qualified as a genuine "microbiocircus" where the microbes are cleverly tamed. *Vibrio alginolyticus,* for example, synthesizes a single sheathed flagella when it develops in a liquid media. Sown on an agar, it develops a crown of cilia, the elements of which are deprived of sheaves. Some authors give a functional interpretation to this difference, considering that a single flagella is enough for mobility in a liquid

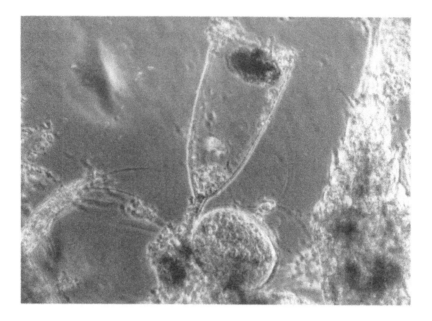

Photo 1.   A true cone of bacteria.

phase, whereas several become necessary on a solid surface. Observations performed on the most varied surfaces in nature confirmed these morphological variations in function with the nature of the substrates and the conditions of existence. Photos 2 and 3 show that they can be considerable, and that a *Salmonella* can take on the aspect of leptospires, rectilinear filaments, or even dwarf forms (J. Brisou and F. Denis, 1978,[252]).

In some sediments, "bio-depositions" can be observed. These are deposits left by mollusks (J. Sonin, 1984). These deposits, biological in origin, and rejected by filtrating mollusks, modify the composition of the sediments considerably, and even the microbiocenoses that benefit from this wealth of food. There are nevertheless materials resistant to all the degradation processes. Alexander qualifies them as "stubborn". Sediments sampled at great depths still reveal a wide range of surprising enzymatic activities (J. Brisou, C. Tysset, R. Moreau, F. Fernex).[245]

Numerous studies have been devoted to the evolution of microbial populations in water: rivers, estuaries, seashores, sources, salty lakes, and even an atoll of the Pacific: Clipperton (P. Niaussat, J. Brisou, and J. Ehrhardt).[246] The results are presented in the summary charts. All the surveys performed on the environment show that the attempts at mathematic modelization concerning microbiocenoses, their dispersion, and their mortality are erroneous from the start because they do not take into account the adherence of bacteria (or other microorganisms) nor their organization in colonies and in aggregates of extreme complexity.

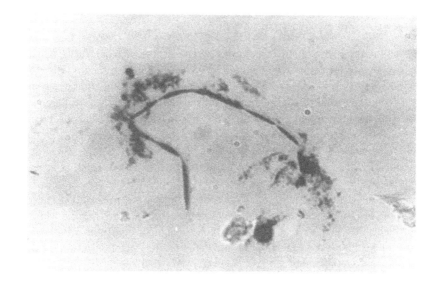

Photo 2.   *Salmonella paratyphi B* after 35 days in natural peptoned (0.5%) seawater-*Leptospira* forms and fragmentation.

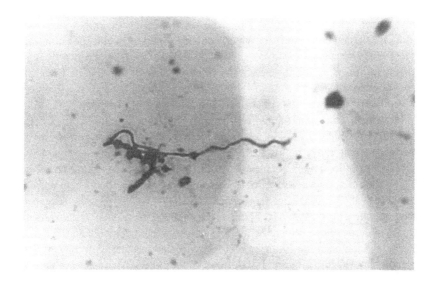

Photo 3.   *Salmonella paratyphi B* after 35 days in natural peptoned (0.5%) seawater-*Leptospira* forms and fragmentation.

5. GAS/LIQUID INTERFACES

All the layers of both fresh and salt water constitute such interfaces, which is also the case for froth, where a gas is enclosed in a bubble. Foam falls into this category as well. Absorption surfaces become significant and have been used for a long time for concentrating substances and microbes. In the past, they were used for isolating tuberculogenic mycobacteria in spit and pus. Foam is especially loaded in microorganisms, which does not decrease interest for the swimming areas at beaches. We had the opportunity of isolating a bacillus responsible for tetanus from one of these foams in a sample taken at a very popular beach.

The interface separating air from water forms a microlayer that stocks hydrophobic substances and a mixture of debris, the very low specific weight of which is lower than that of water. Such substances constitute an obvious food intake for microorganisms. Lemich studied the captation power of the bubbles, thus confirming the conventional data that, upon arriving at the surface, these bubbles can persist, constituting froth and foam, or burst and release the products they have captured. (See Photo 4.)

6. NONMISCIBLE LIQUID INTERFACES

These interfaces separate two nonmiscible liquids. This is a possibility that has become relatively frequent since the multiplication of pollution by oil products. Natural encounters for which humans cannot be held responsible can nevertheless not be excluded. In any case, the interface takes on two dimensions here: it acts as an interface in the strictest sense of the term by favoring the deposit of substances contained in the aqueous phase, but also as a function of its intrinsic qualities. Its composition and nature are such that it has nutritive qualities from which microorganisms benefit. This is the case for hydrocarbons and numerous natural oils. In relation to these oil/water interfaces, bacteria, yeasts, and fungi have highly variable behaviors.

For example, certain bacteria leave the aqueous phase to settle comfortably in the oil and proliferate there. This is the case for acid-resistant germs. Kjellberg et al.[247] showed that *Acholeplasma laidlawii* and *Serratia marinorubra* penetrate into a monomolecular lipidic layer, whereas *Pseudomonas fluorescens* remain underneath the film. Other bacteria are implanted perpendicularly to the interface. Such behavior is attributed to a hydrophobic polarity of the cells or, on the contrary, to the hydrophilia of one of the poles.

In this field of nonmiscible liquids, borders separating liquids apparently of the same nature, but differentiated by temperature and salinity, can be observed. Such interfaces define thermoclines and haloclines on the one hand and the encounter of fresh water and salt water in the estuaries on the other. (See Figure 45.)

These interfaces offer many opportunities for concentrating varied materials and microorganisms (C. ZoBell),[243] J. Brisou et al.[252]). There are direct

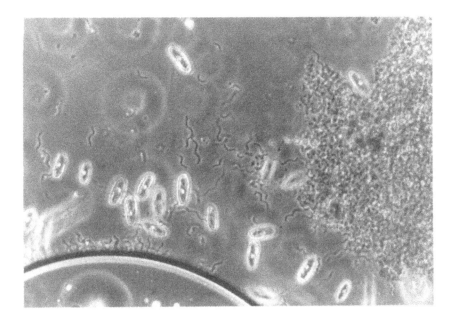

Photo 4. Bacteria and ciliated protozoa on a bubble interface in sea foam (polluted coastal area).

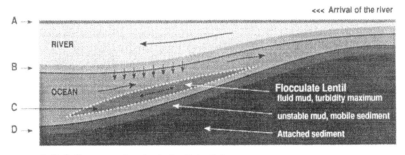

A, B, C, D = zones of maximal microbial activity

Figure 45. Schematic cross-section of an estuarine zone. These very simplified schemas obviously do not take into account hydraulic variations which in an estuary are substantial, but they correspond to very acceptable general locations. In some estuaries, the turbidity maximum can cover from 20 to 25 km at each tide. All the organic or mineral suspended particles constituting the seston are loaded with microbes (bacteria, viruses, yeasts, lower fungi). All these microbes are ensured long life but totally evade the enumerations made using conventional methods.

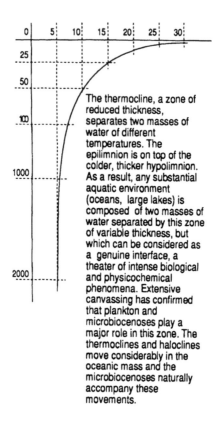

The thermocline, a zone of reduced thickness, separates two masses of water of different temperatures. The epilimnion is on top of the colder, thicker hypolimnion. As a result, any substantial aquatic environment (oceans, large lakes) is composed of two masses of water separated by this zone of variable thickness, but which can be considered as a genuine interface, a theater of intense biological and physicochemical phenomena. Extensive canvassing has confirmed that plankton and microbiocenoses play a major role in this zone. The thermoclines and haloclines move considerably in the oceanic mass and the microbiocenoses naturally accompany these movements.

Figure 46.   Graph of thermocline.

links between the distribution of bacteria and the dynamics of oceans. The situations are even more complex when estuaries are studied. Relatively warm fresh water does not immediately mix with denser and generally colder ocean salt water. In 1978, while prospecting the Loire estuary in France, Frenel[248] confirmed the capital role of clays in the distribution and survival of microorganisms, notably of bacteria of fecal origin. He also insists on the return to "circulation" of muds and bacteria during winter storms. Monitoring the evolution of these populations, he shows the considerable role of the turbidity maximum, an inexhaustable stock of microorganisms of all types.

Irrespective of the type, structure, and composition, an interface is covered in less than an hour with organic and mineral materials that will mandatorily provide shelter for microbial populations. The latter organize, colonize, and form biofilms. The surfaces come to *life*. (See Figures 46–47.)

7. MICROBIAL ACTIVITY ON INTERFACES

The observation of microbiocenoses populating interfaces in an aquatic environment, as well as any other environment, has been the topic of a

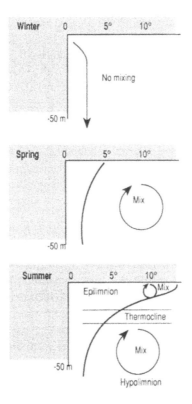

Figure 47.   Locations of the thermoclines in shallow seas.

substantial number of studies. For the time being, we will only discuss the first ones. In a liquid, often oligotrophic medium, traces of energetic elements tend to concentrate on all available interfaces.

Heukelekian and Heller[249] monitored the evolution of *Escherichia coli* in a liquid in which glass marbles were introduced. The oligotrophic, aqueous phase could not enable the growth of bacteria but the latter, once attached to the glass marbles, enriched in food and developed actively.

Jannasch[310] confirmed the colonization of chitin particles by *B. subtilis* in a liquid medium very poor in substrates. As chitin is by itself a consumable, biodegradable substrate, the experiment is less demonstrative than the first one in which the substrate is inert. In nature, microbes rapidly colonize all media — sand, gravel, sediment and all suspended matter — while abiding by these mechanisms. The biosynthesis of the adhesins is confirmed in the presence of chloramphenicol, the inhibitor role of which is known. As the germs cannot ensure biosynthesis, adhesins no longer adhere to the interfaces. The colonies settled on the substrates generally get along well, but in a few environments the living conditions and food deficiencies can cause competition.

Microbes are present everywhere we look. Microbial life and organic matter are closely related. A microbiocenose will be more rich and diversified the more abundant and heterogenous the organic matter. The selective character of sediments, like that of soils, is a function of the grain size that governs the distribution of nutritive elements. All interfaces, irrespective of their dimensions, are the seat of bacterial colonizations and attractions. The walls of a basic tank of water, hydrocarbons, or oil become the seat of an often very intense microbial activity in just a few hours. The interface constituted of two nonmiscible liquids rapidly becomes the theatre of an intense bacterial activity altering both the hydrocarbon and the metal walls of the tanks.

Although marine sediments are, like soils, rich in a variety of materials, a special place is attributed here to humic substances mentioned previously. They are complexes to which Stuermer and Harvey devoted a study in 1974. This marine humus is indeed richer in hydrogen and nitrogen than the earth, whereas its level of oxygen is lower.

Fulvic and humic acids constitute excellent sites for bacteria and all other microorganisms. Rich in carboxyls, hydroxyphenols, valine, proline, and leucine, they constitute a substantial food reserve for these microscopic beings. Plankton contributes to a major part of the supply of sediments in humic substances, either due to their mortality or due to their dejections. The remainder is partially alluvial in origin. The biosynthesis of the chemical and structural elements of adherence are closely related to the wealth of the medium, so the importance of humic intakes in the progress of all these processes can be understood. As these acids are, moreover, very powerful cation chelators, notably of heavy metals, they participate in the safeguarding of living beings in possibly contaminated environments. Chelated metals buried in the humus do indeed become much less dangerous than in the state of soluble salts.

8. ROLE OF PLANKTON IN AQUATIC ENVIRONMENTS

No one questions the important role that plankton plays in the stocking and transportation of microorganisms in water. Phytoplankton preserves them for a certain period of time on the surface and at reduced depths. Zooplankton absorbs them and drags them into its savings. With Rigomier,[250] we devoted a certain number of research studies to this aspect of marine life, then with Blavier,[251] Denis,[252] and Moreau,[253,254] in the Atlantic Ocean and the Mediterranean Sea, even at great depths. All the coastal samples (as well as those taken in the open sea) were carriers of Enterobacteria: *E. coli, Citrobacter, Edwardsiella, Proteus vulgaris, Morganella, Rettgerella*, etc., without counting numerous other bacteria common to the environment: *Achromobacter, Flavobactrium, Xanthomonas, Cocci, Bacillus*, etc.

On some samples of zooplankton, Rigomier[255] succeeded in isolating 14 different bacterial species by dissecting the intestine. Plankton is therefore an important reserve of often undesirable bacteria. The samples caught in the estuaries and near the coasts were 100% carriers of putrid anaerobic bacteria (*Clostridium*), very often *E. coli*, and *St. faecalis*. As shellfish farms are most often set up in the estuary zones, the mollusks themselves feeding off of plankton, it should be understood that their analysis is much more important than that of water. Plankton and sediment are the essential elements of these environments.

By microenzymatic procedures, we confirmed the substantial activity of bacteria in marine plankton. The pollution of an estuary necessarily contaminates the plankton, sediments, and everything that depends on it. Dienert and Guillerd[256] (1940), Carlucci and Pramer[257] (1960), Oppenheimer[258,259] (1960), Brisou et al.[260,261] (1965–75), Rigomier[262] (1979), Campello et al.[263] (1963), Blavier[251] (1971) and many more authors participated in these investigations on coastal plankton and high sea samples. All the results confirm the large bacterial load and the oceanic peregrinations of these pollutants.

Kaneko and Colwell[264] (1974) monitored the evolution of *V. parahaemolyticus* in Californian waters. This bacteria is possibly pathogenic for humans. Absent from coastal waters, the authors found it perfectly viable in the zooplankton sampled in the open sea. The authors did not hesitate to write that it was "the most important determining factor in nature". The vertical breakdown (jagged line) of bacteria in water abides by the laws of plankton distribution submitted to the chemical and physical conditions of the environment — to light, temperatures, currents, salinity, etc.

The daytime decrease of the number of bacteria counted in the euphotic zones, is often attributed to the nocive effects of ultraviolet rays. It is more accurate to see the influence of zooplankton migrations, loaded with bacteria, that dive towards the depths in fear of the sun and resurface to spend the night. Vinogradova[265] specifically monitored these rhythms of the various species of zooplankton. We also participated in this type of research off the Dakar coast and, thanks to the cooperation of Curcier,[266] by performing direct counts on colored Millipore filters.

These concepts lead quite naturally to the mollusks which feed off of plankton. They too are quite rich in bacteria and in viruses, but it does not appear useful to insist on this particularity. The images reproduced in Plates 1–3 and in Photos 5–11 should be enough to confirm this fact.

In the chapter devoted to experimentation, we will show the speed at which certain bacteria attach to shellfish.

9. POLYMORPHISM OF BACTERIA IN NATURE, AGGLOMERATES

Under the pressure of the surrounding conditions, microorganisms offer images sometimes extremely far from what we are used to seeing in the laboratory. Such bacteria, normally present in the form of a regular stick, become a genuine leptospire; others, on the contrary, reduce their size, become hard to distinguish and so fine that they pass through the filters. "Dwarf" forms were described by Kauffmann.[35] We were personally interested in these

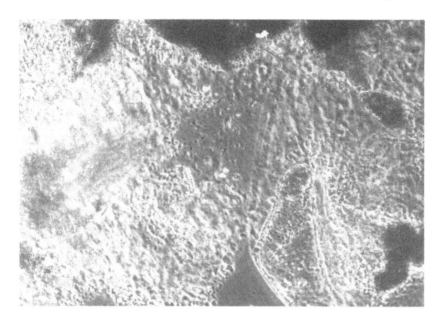

Photo 5. Zooplankton covered by bacteria.

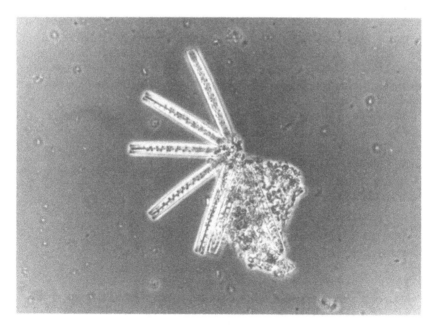

Photo 6. Bacterial agglomerate on phytoplankton.

Photo 7. *Micrococcus* in marine sediments.

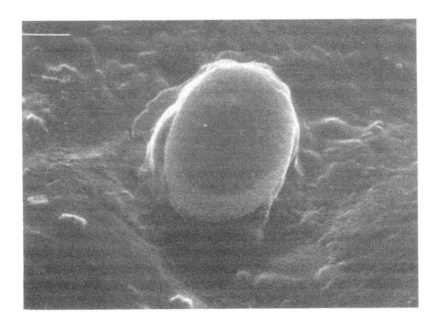

Photo 8. Adhesins on wild bacteria.

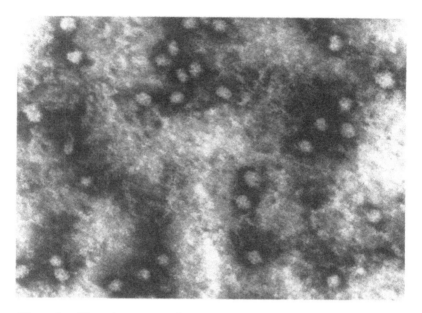

Photo 9.   Virus in oysters, *Crassostrea crassa*. (Photo by F. A. Denis)

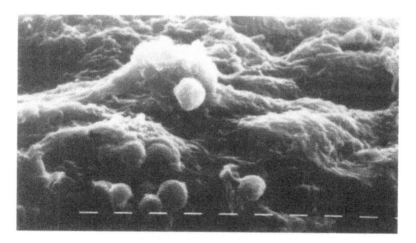

Photo 10.   Bacteria in *Crassostrea crassa*.

transformed, hard-to-evidence bacteria. Often present in marine environments, sediments, and shells, we qualified them as "masked bacteria" (J. Brisou,[4] 1969). Certain authors preferred calling them "debilitated forms". They are relatively easy to evidence thanks to certain tricks of classic cultures. These germs, considered "dead" or "disappeared", are in reality still present and alive. One just needs to know the appropriate means of "releasing" them in order to "unmask" them.

Photo 11.   Extraordinary concentration of bacteria in a Mediterranean mussel, *Mytilus galloprovincialis* (probably 1/4 of its weight).

Microorganisms are all organized in communities, colonizing the interfaces as we have just seen. To make these colonizations a success, they often form very large "aggregates" that are genuine carpets rich in proteins, lectins, and polysaccharides, and surrounded by an ionic aura. Capsules, microcapsules, polysaccharide layers, fimbriae, glycocalyx, etc., cooperate in the organization of these clusters. Adhesins all take part in these mechanisms. In *Streptococci*, for example, the union of cells is provided by interparietal bridges perfectly evidenced by electron microscopy.

These bonds are broken by lysozyme, as shown by Webb[268] (1948) in the majority of Gram-positive bacteria. They are also broken by other enzymes such as hyaluronidase and polysaccharases. In *Sarcina ventriculi*, the cement appears to be cellulosic in nature, according to Canale Parola et al. (1961). Other Gram-positive germs also unite thanks to polysaccharides or to parietal

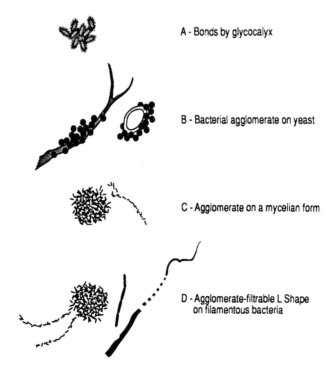

A - Bonds by glycocalyx

B - Bacterial agglomerate on yeast

C - Agglomerate on a mycelian form

D - Agglomerate-filtrable L Shape on filamentous bacteria

Figure 48. **Behavior of microbes with agglomerates. Schemas inspired by personal photographic documents.**

glucosaminopeptides. In other circumstances, fimbriae, lectins, capsules, and microcapsules contribute to this formation of agglomerates.

The work by Hauduroy established the affinities between *Mycobacterium tuberculosis* and *Candida albicans*. The presence of this yeast on a Loewenstein-type culture medium activates the growth of the agent of the tuberculosis. We photographed this particularity showing the formation of a rosace of bacteria around a yeast. (See Figure 48 and Photos 12–18.)

To summarize, bacteria prefer gregariousness to isolation. They only live well in communities. They invade living interfaces as well as inert ones. H. Pearl[269] (1980) insisted a great deal on this behavior. According to Pearl, the variability of the soluble particles of substrates over time and space constitutes the dominating factor in the propagation of the attached bacteria. The importance of agglomerates in aquatic environments is too often underestimated. We openly subscribe to this opinion which has been widely confirmed by observations. We hope that the photographs published here will be convincing. They were all taken in "real life" situations during patient "bacteria hunts".

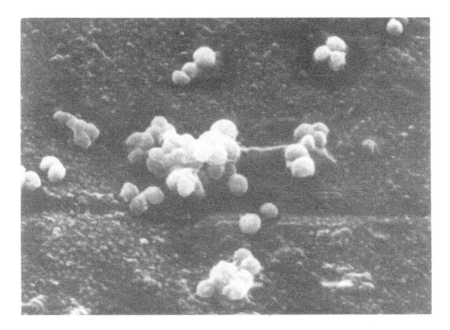

Photo 12.   Filtrate on Millipore filter of sedimentary dilution.

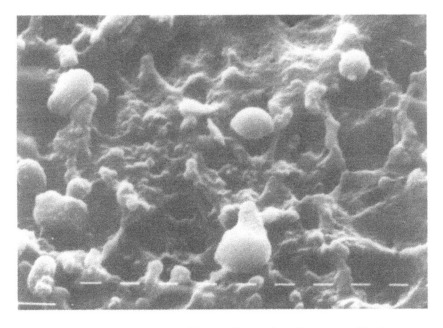

Photo 13.   Filtrate on Millipore filter of sedimentary dilution.

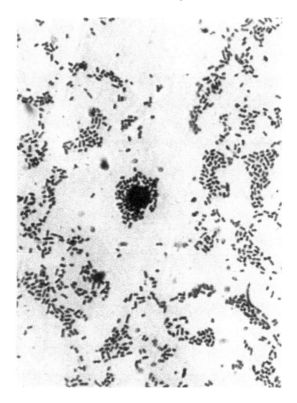

Photo 14.   Bacterial aggregate on yeast.

10. INERT SURFACES

It is hard to define an "inert" surface; it seems more appropriate to speak in terms of inorganic interfaces or surfaces: metals, glass, rock, plastics, polystyrene, etc., exposed to air, submerged in various liquids, or buried in soils or sediments. In all but a few exceptions, these surfaces are rapidly colonized by microorganisms sometimes posing serious problems for industry and hygiene in general, as demonstrated by a few quick examples.

It will be shown in the end that an interface is practically never "inert". Contamination is unfortunately frequent of common surgical instruments, including metals, plastics, glass, rubber, syringes, and catheters, which all become good shelters for undesirable bacteria and lead to so-called "nosocomial" or iatrogenic infections, notably at the time of long infusions.

Along with our colleagues, we had the opportunity of recording these accidents in our programs: A. Nivet[270] made a record of infections caused by soiled catheters (1971) and F. Denis et al.[271] monitored the proliferation of bacteria in artificial incubators (1972). *Pseudomonas* of limited demands were relatively frequent. These observations were published in most hospitals.

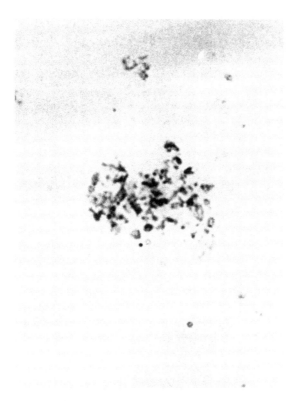

Photo 15. Bacteria on millipore filter, with a supposed yield of only a small colony on culture media.

Plastic- and rubber-based instruments are especially receptive to bacteria and other microorganisms.

These contaminations concern industry in the same way, notably the food industry and, more particularly, the dairy industry. The study published by Driessen et al.[272] in 1984 concerns the adherence of *Streptococci* thermoresistant to the stainless steels composing the tanks used in dairy farms. Volumes of 500-ml of raw milk and pasteurized milk were planted with a strain of *S. thermophilus B* at the dose of 10^6 cells/ml. Sheets of stainless steel were hung in these mixtures. After an incubation of 3 hours at 35°, 40°, 45°, and 50°C, the number of bacteria attached to the metal sheets was approximately equivalent in the raw milk and the pasteurized milk! A treatment of the steel by calcium phosphate reduced the adhesion by 50%. When the experiment was continued in a continuous stream of milk simulating the conditions of pasteurization, the adhesion rates were identical for the first hour whether the milk was pasteurized or not. However, after 2 and 4 hours of observation, the colonization reached 10 million cells/cm² in the presence of pasteurized milk as opposed to only 700,000 with the raw milk. The treatment of the sheets with calcium

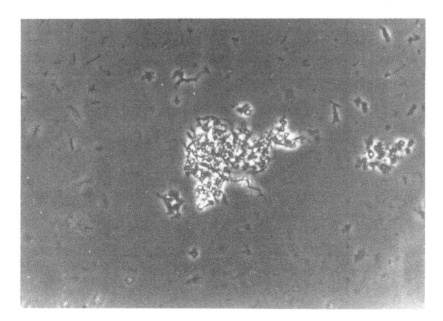

Photo 16.   Bacterial agglomerate in seawater.

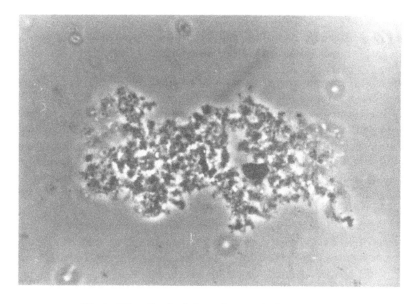

Photo 17.   Bacterial agglomerate in seawater.

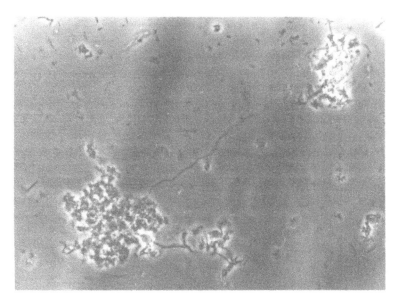

Photo 18.   Bacterial agglomerate in seawater.

phosphate had no noticeable effect in the presence of the raw milk, whereas it reduced the adherence to 4.2 million/cm^2 in the pasteurized milk. The authors suggested the intervention of a thermolabile factor, an inhibitor, that was opposed to the adherence of bacteria on the metal in the raw milk. It can also be supposed that numerous microorganisms present in all the raw milks begin to proliferate after the 4th hour. They enter into a logarithmic phase, the most favorable to the elaboration of enzymes and other factors capable of "releasing" the already attached germs or of opposing these adherences by mechanisms known to be multiple.

We observed the same phenomenon in seawater aquariums experimentally contaminated with 32P-labeled bacteria, as specified in Part III. On the other hand, the presence of calcium phosphate modifies the potential of the sheets and therefore the ionic conditions of these interfaces and their receptivity to adhesion molecules. L. Stone and E. Zottola[273] (1985) confirmed this attachment by monitoring the behavior of *Pseudomonas fragi* on stainless steel sheets in the presence of skimmed milk incubated at 4° and at 25°. Adherence occurred regardless of the conditions of rest or agitation of the liquid. This attachment was provided by fibers in 0.50 hour at 25° and in 2 hours at 4°. It increased as the growth increased. Bioluminescent bacteria constitute a material of choice for monitoring these phenomena. S.P. Deneyer et al.[274] (1991) observed the dynamics of attachment of a bioluminescent phenotype of *E. coli* in this way on polystyrene sheets.

The same technique was used by Mittelman et al.[275] who experimented with a strain of *Pseudomonas fluorescens*, a carrier of the operon specific to *Vibrio fischeri* which is normally bioluminescent. The adherence was estimated at 2×10^5 cells/cm² of surface. There was a correlation between the synthesis of lipids on the biofilm and the attachment of bacteria, moreover, controlled by 14C, in the form of acetate incorporated into the experimental system. After coloration with orange acridin, the authors enumerated between 100,000 and 10,000,000 cells/cm². These values are worthwhile as they provide an idea of the differences recorded between the number of germs in a liquid and the quantity attached on the walls of a recipient. We already evoked these problems for sand and coastal waters.

D. Allison[276] (1993) recalled the commercial and industrial consequences of bacterial deposits on submerged surfaces and insisted on the essential role of exopolysaccharides produced on the interfaces. Adhesins play a crucial role in the evolution and maintenance of biofilms. According to B. Bendurger et al.[277] (1993), the long chains of mycolic acids formed by the corynebacteria, *Rhodcoccus*, mycobacteria, *Arthrobacter,* and other coryneforms containing hydrophobic type B peptidoglycans facilitate adherence on the surfaces of teflon and glass.

Furthermore, it has been observed that although ferrous ions inhibit the selective adhesion of *Thiobacillus ferrooxidans* by competition with pyrite, ferrous ions exert practically no influences on the phenomenon. Bacteria recognize the ions reduced in the mineral and adhere strongly to chalcopyrite and pyrite (N. Oh Mura et al.,[278] 1993). This same *Thiobacillus ferrooxydans* also interested P. Devasia et al.[279] the same year. The germ cultivated in the presence of sulfide, pyrite, and chalcopyrite becomes much more hydrophobic than if it were developed in the presence of ferrous ions. As the antigenic structures are modified, the germ no longer reacts with the specific antiserums corresponding to the strain raised in a ferrous medium. A new protein appears on the surface and facilitates the adherence to mineral substrates. Bacteria originating in ferrous media do not have these anchorage fibers.

On this topic, it is advantageous to consult the work devoted to the deterioration of materials by bacteria, notably *Microbial Aspects of the Deterioration of Materials* by D. Loveleck and R. Gilbert (1975), as well as those works dealing with the metabolism of *Thiobacillus,* prototypes of bacteria developing in extreme environments. (Konetzka,[280] 1977; Murr,[281] 1978).

For A. Brouillard-Delattre[282] (1993), biofilms become an industrial reality. The author insists on the importance of the massive colonizations on the equipment commonly used in the dairy industry: stainless steel, teflon, seals, rubber, etc. Photographs taken using electron microscopy illustrate his point of view.

Dunne et al.[283] (1993) observed the attachment mode of *Staphylococcus epidermidis* on plastic surfaces. Apparently, the germ previously covered with a fibronectin of human origin adheres less easily than bare bacteria. The same behavior is observed in bacteria coated with fibronectin linked with heparin and gelatin. The strains in the negative phase do not appear effected by these treatments. The same phenomena are found on surgical instruments as well. B. Carpentier and O. Cerf[284] (1993) analyzed the consequences of these biofilms

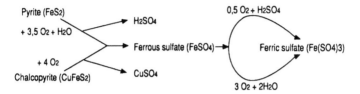

Figure 49. Oxidation of sulfide by *Thiobacillus ferrooxidans*.

in industry and attempted to draw some practical conclusions on a sanitary or technical level and, more specifically, as far as instrument maintenance is concerned. Numerous undesirable bacteria participate in dirtying up surfaces of all kinds: *Pseudomonas, Salmonella, Yersinia enterocolitica,* various *Streptococci, S. aureus, Listeria,* etc., are frequently identified. We had the opportunity of monitoring the contamination of this type of equipment by a strain of proteolytic *Staphylococci* in a cheese factory (unpublished) where it was the source of a substantial drop in the factory's turnover.

Listeria monocytogenes, responsible for small epidemics, still draws attention. K. Sasahara et al.[285] (1993) studied the behavior of this bacteria and *Pseudomonas* on glass slides submitted to a continuous stream. The observations were performed under epifluorescence and electron microscopy, sometimes with pure cultures of each of the germs and sometimes with a mixture of both. *Listeria monocytogenes* is attached in dispersed cells, whereas *Ps. fragi* forms a genuine rug of flowing cells. The mixture of both cultures causes the massive production of an exopolymer promoting the adherence of *Listeria*. It is suggested that this exopolymer modifies the surface ionic charges, activates the mobility of flagella, and thus facilitates adherence. In any case, the behavior of germs is different depending on whether they are from pure cultures or whether they are associated.

A phenomenon very close to beneficial association was observed by Beech et al.[286] in 1991. These authors monitored the formation of a biofilm on stainless steel sheets in the presence of milk and strains of *Desulfovibrio desulfuricans* and *Pseudomonas fluorescens*. A higher production of polysaccharides in the presence of the two associated bacteria was observed. Mannose dominated all of these extracellular coats.

Thus, without having to multiply the examples, we can assert that all living or inert interfaces populated by bacteria or other microorganisms constitute biofilms, active biomasses endowed with singular properties related to the microbes themselves and to the products of their metabolism. Their consequences in fields ranging from medicine to industry have been put forward.

11. CONCLUSION

The life of microbes in the wild places the observer very far from what he or she can observe *in vitro* in the laboratory with pure strains raised in defined media and incubated at regulated temperatures. The adjective "wild" that we have been using for several years is by no means negative or humorous; it corresponds to an ecological reality that is easy to comprehend. A *Salmonella*, an *E. coli*, a *Listeria*, a *Clostridium*, etc., or other "room" bacteria have nothing in common with these very bacteria in nature mixed with disparate crowds in water, soils, food, in the intestine, in the heart of sediments, and lost in the immensity of the biosphere where they are never independent.

In a recent article, Szmelcman[287] accurately analyzed the complex mechanisms enabling Gram-negative bacteria to live in nature, irrespective of the environmental conditions. According to the author, these are properties "acquired during evolution, where perceptions, transductions of information and signals are combined". Do bacteria have a conscience? We have a right to wonder since they perceive and react in their best interest despite their primitive appearance (1994). As all living beings, they are constantly adapting to survive and last. Such is the meaning of microbial ecology.

Knowing the formation mechanism of biofilms, their structure, and the chemical composition of their elements (for the most part), we undertook the development of enzymatic release techniques that were as rational as possible. These techniques are aimed either at destroying the substrate of the biofilm or at releasing a specific bacterial species worthy of interest.

For every soil, sediment, metal surface, fabric, or pathological product, there are as many biofilms. The inert character of an interface is only conceivable after sterilization. We summarized the consequences of contaminations and pointed out the participation in the pathology of usually commonplace "opportunist" bacteria which adhere to previously sterilized inorganic surfaces. These substrates, once colonized, become responsible for sometimes very serious diseases. In these conditions, it goes without saying that the industrial circuits of manufacturing and distribution of foodstuffs must be supervised with extreme caution to avoid accidental contaminations.

These observations, collected for over a century, are a plea for the multiplication of team work, as advised by R. Moreau.[288] Microbioecology, epistemologically speaking, is only conceivable by associating research conducted by geologists, oceanographers, physicists, chemists, microbiologists, agronomists, industrialists, etc. This will be the topic of Part III.

PART III
Enzymatic Release Techniques of Bacteria

1   BASIC TECHNIQUES

1. INTRODUCTION

The principle of enzymatic release of microorganisms from their shelters to identify them and make an inventory was suggested in 1979 (J. Brisou).[241] Experiments had started in 1978, at a time when the biochemical mechanisms of bacterial attachment were increasingly well known, as well as the anchorage organs, grouped under the term adhesins. This data quite naturally led to an attempt at breaking these genuine "moorings" enabling the bacteria to attach to interfaces. Specific hydrolases were likely to be of help. This idea was put forward in 1965 by Skujins et al.[290] who had succeeded in destroying the cell walls of fungi thanks to a chitinase and an alpha-3-glucanase.

S. Hosegawa and J. Nordin[291] confirmed this observation by using an alpha-D-glucanase from Trichoderma viride, a usual host of soils and producer of a gliotoxin. O. deVries et al.[292] continued similar research with an enzyme of the same origin (T. viride) which this time was used for "releasing" the protoplasts of *Schizophylum commune*. They showed that the enzyme acted essentially on R-glucans, laminarine, and slightly on starch. This enzyme was identified with an alpha-3-glucanase very close, but not identical to, that encountered in the gastric juices of certain snakes. It releases the protoplasmic spheres by breaking the walls and it acts at pH 5.8 after release of the protoplasts. A regeneration of approximately 10% of the walls destroyed by hydrolase can be seen.

Such observations only confirm the well-foundedness of our undertakings, but our goal is clearly different from the one that inspired the authors just mentioned. The reader should understand that our programs are essentially focused on the microbism of living beings and microbial ecology so as to draw up inventories of bacterial populations inhabiting the most wide-ranging environments.

The conventional methods used for bacteriological analysis of water, foodstuffs, pathological products, sand and sediments, etc., only take into account the rarely free, unattached bacteria most often present in agglomerates. The germs attached to interfaces of all sorts, whether inert or alive, are not detected by this canvassing. For example, we noted considerable differences between an analysis of water from a very popular swimming pool and the analysis of products obtained by scraping the walls of this swimming pool. The deposits were particularly rich in *Pseudomonas aeruginosa* and *Streptococcus faecalis,* which had not been detected by the routine analysis of the water.

The results are, from the start, marred by mistakes for another reason: in the usual conditions of analysis, a bacterial colony on a solid medium never corresponds to one bacterium but rather to 50, to 100, or even more. It can therefore be asserted that the methods that have been used for almost a century have only provided questionable results, as we will have the opportunity to demonstrate shortly.

Enzymatic release offers:

- The attack of host sites or substrates
- The destruction of adhesins characterizing living tissue and bacteria.

2. ATTACK OF HOST SITES OR SUBSTRATES

These formation surfaces of biofilms are either inorganic, organic, living, altered, or dead.

2.1. Inorganic or Inert Surfaces

As a whole, they are not very sensitive to hydrolases. In general, they are metals, silica-based materials, glass, rocks such as granite, marble, or other mineral elements, but also synthetic coverings falling within the realm of organic chemistry and often "stubborn" according to Alexander.[2] Plastics, polystyrenes, and other substances of this type are especially brought to mind.

2.2. Organic Surfaces

Organic surfaces are, on the contrary, quite often hydrolyzed at least partially and sometimes totally by judiciously chosen enzymes. This is the case for the cellulose- or chitin-based substrates which are excellent captors of microorganisms, a variety of proteins, dead or altered tissues, vegetables or meats, milk, butter, etc. For years, numerous processes have been used that have followed technical progress for getting rid of microorganisms contaminating interfaces. In this field, everything depends on the goals to be reached. Sterilization remains the simplest, most widely used means; dry heat, high-temperature steam (Uperization), autoclaves, gamma rays, and antiseptics do not need to be developed here. Our goal is limited to the (as complete as possible) study of bacterial populations in a specific environment or on an organism. Conventional methods are based on dilutions, grinding, and often brutal or traumatic mechanical dispersions. Tensioactive and disorbant methods are also suggested which are sometimes harmful to bacterial life.

Thanks to a better knowledge and more rational comprehension of bacterial attachment, and also to data input from the biochemistry of adhesins, parietal structures, and cellular surfaces, using enzymes capable of destroying these attachment sytems was the logical option. This was the point of departure for specific physiological techniques and, as a whole, not particularly traumatic for bacteria, with a few rare exceptions since there are bactericide or at least bacteriostatic enzymes.

We remained loyal to a few enumerations while recognizing their weaknesses pointed out in the last century by Duclaux[1] and Winogradsky.[21] A colony only corresponds to a single cell when using very particular and meticulous methods of cloning, isolation of cells separated by micromanipulations and microcultures, which is not usual. The colonies appearing on solid media of common use therefore result from numerous cells and agglomerates. Their number depends on the culture media chosen and the incubation temperatures.

We contributed substantially to this type of research with our colleagues or alone. With Y. de Rautlin de la Roy,[260] we cultivated many samples of seawater taken in the North Atlantic Ocean up to 4000 meters in depth, thanks to the meteorological ships France I and France II.

Each sampling was cultivated simultaneously on freshwater media and media with a natural seawater base. Each one was systematically incubated at 18 and 37°C (see chart above). For other biotopes, we used from four to eight different media with natural seawater bases and prepared with freshwater. The incubations were performed at 4, 18, 37, 45, and 55°C. In some cases, we used 15, 20, or 25% hypersalted media. The findings are summarized in the charts, simple examples among many others which have been the subject of doctoral theses in medicine or science. In 1946, Zo Bell[243] only used a single media. He incubated the seawater cultures at temperatures ranging from 4 to 37°C (see chart), but extended the observation time.

2.3. Analysis of North Atlantic Seawater

	Culture		Medium	
	Salt	Fresh	Salt	Fresh
Depth	**18°**	**37°**	**18°**	**37°**
0	2475	5222	3795	204
250	750	3600	1800	1500
1000	3600	0	0	10,500
1500	414	36	221	66
2500	+++	2310	++++	190
4000	7200	3600	3000	3600
4000	18,000	0	8400	0

Data from Brisou, J. and de Rautlin de la Roy, Y. *C.R. Soc. Biol.* 159, 1454, 1965.

A few values were provided by ZoBell in 1949.[243]

Incubation in days	4°	12°	18°	22°	25°	30°	37°
1	0	18	30	36	41	48	8
10	9	67	91	96	84	71	13
18	20	97	100	95	84	63	

Note: The figures represent the number of colonies/100 ml of water. +++ = complete invasion of nutritive agar.

3. USEABLE ENZYMES

During former experiments, we studied the detergent power of papain that we considered as a "genuine biological knife" (A. Rigaud, J. Brisou, R. Babin, 1952).[296] Associated with antibiotics, it was at the origin of a therapeutic treatment that has proved itself. Nevertheless, as far as current problems are concerned, we limited the experimentation to lipase and several polysaccharases easily found on the market. The choice is extremely vast since the international nomenclature gives a list of 74. As the industrial offer is less generous, we will only discuss marketed enzymes here. They belong to the group of hydrolases E.C.3.1. and 3.2. of the official code.

E.C.3.1.1.3 Lipase It is a triacylglycerol lipase catalyzing the following reaction:

$$\text{Triglycerol} + H_2 0 = \text{Diglyceride} + \text{fatty acid}$$

On the market, there are a dozen varieties of lipases extracted, for example, from wheat germ, pig pancreas, *Candida cylindraceae, Geotrichum candidum,* and *Rhizopus arrhinus.* One unit hydrolyzes 1 mEq of fatty acid of a triglyceride in 1 hour at pH 7 and at 37°C. The most commonly used titration medium is triacetine. We used this enzyme at various times with noteworthy results. For example, it releases *S. faecalis* from a beach sand after a contact of 20 hours at 28°C whereas the control did not generate any colonies of the germ. On a Sabouraud media, we counted 470 colonies produced by a micrococcus for an inoculum of 50 ml. The same lipase released *Candida zelanoïdes* attached experimentally on kaolin.

E.C.3.2.1.1 Alpha-amylase (diastase, ptyaline, glycogenase) The official name of this enzyme is: 1,4-alpha-D-glucane-glucanehydrolase. It acts on starch, glycogen, similar polysaccharides, and oligosaccharides of the group. Alpha structure reducer residues are released. Numerous preparations of this enzyme are available on the market. One unit provides 1 mg of maltose using starch in 3 min at pH 6.9 and at 20°C. As a result, products from pig pancreas, *Bacillus subtilis,* or even *Aspergillus orizae* are available. The activities oscillate between a few hundred units to 1500 U/mg of protein.

E.C.3.2.1.2 Beta-amylase or 1,4-alpha-D-glucane-maltohydrolase

It catalyzes the hydrolysis of alpha-1,4-glucan bonds of polysaccharides and releases successive units of maltose from the nonreducer end of the chains. Not only are the 1,6-glucan bonds resistant to the attack, but they constitute an obstacle to the activity of the enzyme. The smallest molecule hydrolyzed by the enzyme is maltotetraose. The unit corresponds to the release of 1 mg of maltose in 3 min at pH 4.8 and at 20°C. The product marketed comes from the sweet potato or from barley with activities ranging from 750 to 1000 U/mg of protein.

E.C.3.2.1.3 Amyloglucosidase or glucosamylase

It is an exo-1,4-alpha-glucosidase alpha-1,4-glucane hydrolase. It hydrolyzes alpha-1,4-glucan bonds of polysaccharides by freeing successive units of glucose coming from the nonreducer ends of the chains. Maltose, amylopectin, and amylose are entirely converted in glucose. These substrates remain the most specific of the enzyme the unit of which provides 1 mg of glucose from starch in 3 min at 55°C and pH 4.5. The beta-amylase also acts on 1,6 and 1,4 bonds of certain polysaccharides.

E.C.3.2.1.4 Cellulase or 1,4-beta glucan

Officially, it is a 1,4-(1,3-1,4)-beta-D-glucan 4-glucan hydrolase. One unit frees 1 µmol of glucose from cellulose in an hour at pH 5.0 and at 37°C (we recommend 2 h of incubation). The cellulases on the market come from *Aspergillus niger* or *Trichoderma viride*. In general, they titer from 0.1 to 1 U/mg of solid, which is low. The enzyme which comes from *Myrothecium verrucaria* hydrolyzes lichenin, xylane, glucomannane, polysaccharide of Crown Gall, lutean, laminarine (which is valuable for the release of *Pseudomonas*), methyl beta-cellobiosis, carboxymethylcellulose, and sorbitol-beta-cellobioside. *Trichoderma viride* is also a powerful producer of cellulase. The wide spectrum of activity of this cellulase is assessed in function with its origin. The commercial preparations are unfortunately not very strong (0.5 to 1 U/mg).

E.C.3.2.1.11 Dextranase or alpha-D-1,6-glucane hydrolase

One unit of this enzyme releases 1 µmol of isomaltose per minute at pH 6.0 and at 37°C. (dextran medium). The commercial enzyme extracted from a nonspecified *Penicillium* is relatively active. The values given range from 15 to 200 U/mg. Certain titer even up to 5000 U. Unfortunately, their extremely high cost limits their use. Certain species of *Cytophaga* have a dextranase.

E.C.3.2.1.14 Chitinase or 1,4-beta-(2-acetamido-deoxy-D-glucoside)-glucanohydrolase

This is more simply chitinodextrase or acetyl-glucosamidase. This hydrolase acts on certain mucopeptides in the same way as *N*-acetylmuramohydrolase (Lysozyme). The chitinase marketed comes from *Streptomyces griseus*. One unit frees 1 mg of *N*-acetyl-*O*-glucosamine of chitin in 48 h at pH 6 and 25°C. Acetylglucosamine is quantified using the 3,5 dinitrosilicylic reagent. The enzyme was also extracted from *Streptomyces antibioticus*. It acts on chitosan, carboxymethylchitin, glycochitin, and chitin sulfate. However, chitin nitrate, cellulose, mucin, hyaluronate, and alginates remain insensitive. We evidenced this chitinase in a certain number of bacteria of marine origin (J. Brisou, C. Tysset et al., 1964).[294]

E.C.3.2.1.15 Pectinase or polygalacturonase   Poly-alpha-1,4-galacturonide glucanehydrolase hydrolyzes bonds of pectates and other polygalacturonides One marketed preparation comes from *Aspergillus niger* — although other organisms have the enzyme, notably *Saccharomyces fragilis* — one unit frees 1 mmol of galacturonic acid from polygalacturonic acid at pH 4.0 and at 25°C. Oligosaccharides are attacked preferentially on the first glucosidic bond close to the reducer group of the molecule. The isolated enzyme of tomato juice catalyzes the degradation of pectates, tetragalacturonates, and trigalacturonates. Tetragalacturonate gives birth to a galacturonate and a trigalacturonate.

E.C.3.2.1.17 Lysozyme   Also known by the name of muramidase or mucopeptide glucohydrolase, in the international nomenclature it is called mucopeptide-N-acetyl-muramylhydrolase. It has an activity close to that of chitinase (E.C.3.2.1.14) examined above. It does indeed attack the beta 1,4 bonds established between N-acetyl muramic acid (or 2-acetamido-2-deoxy-D-glucose) and the 2-acetamido-2-deoxy-D-glucose residues from the mucopolysaccharides of chitin. The best known source of this enzyme is egg whites. The preparation marketed is in the form of a crystalline white powder widely used in therapeutics. In particular, it is known for its marked bactericide properties for Gram-positive germs and its action on certain Gram-negative bacteria. Lysozyme has not only served as a topic for articles, theses, and papers, but also for entire volumes ever since its discovery by A. Flemming in 1922 (*Proc. R. Soc. London*, B93 306, 1992). Destruction of streptococcus walls isolated from patients using lysozyme was confirmed by electron microscopy (J. Brisou, F. Denis, P. Babin, and R. Babin, 1976).[295]

E.C.3.2.1.18 Sialidase or acylmuramyl hydrolase   This well-known enzyme is none other than neuraminidase. It attacks alpha-2,6 terminal bonds uniting the N-acetyl-neuraminic acid and 2-acetamido-2-dioxy-D-galactose residues entering into the structure of a variety of mucopolysaccharides. It is found in the membrane of the chicken embryo, but also in *Vibrio cholerae* and even in the structure of the grippal virus or in *Diplococcus pneumoniae* and *Clostridum perfringens*. One unit frees 1 mmol of N-acetyl neuraminic acid per minute at pH 5 and at 37°C. The quantification substrate is N-aminic N-lactose of submaxillary mucin. A soluble product coming from *Clostridium perfringens* and a form-attached agarose are available on the market. There are also preparations from choleric vibrion cultures. The allantoid membrane of eggs is also rich in this enzyme. The speed of hydrolysis of the substrate is conditioned by the position of the bonds between the sialic acid to the penultiem sugar of the polysaccharide chain. After extended incubation, in the presence of complex substrates, sialic acid is released. The enzyme from the choleric vibrion is shown to be Ca^{2+} dependent, which is not the case for that of *Clostridium*. The animal-origin preparations are much less active than the microbial ones. They only release 20 to 30% of the sialic acid present in the complexes.

E.C.3.2.1.20 Alpha-D-glucosidase or glucoside hydrolase This enzyme is also known under the name of maltase, glucoinvertase, or glucosidosaccharase. It is more accurate to speak in terms of a group of enzymes that hydrolyse oligosaccharides rapidly whereas polysaccharides are clearly less sensitive, even resistant to the action of these hydrolases. The overall reaction is as follows:

alpha-D-glucoside + H_2O = an alcohol + D-glucose

This enzyme hydrolyzes a wide variety of alpha-D-pyranosides, that of *Saccharomyces italicus* attacks the following compounds: phenyl-D-glucoside, furanose, saccharose, beta-methyl-maltoside, maltose, butyl-alpha-D-glucoside, and ethyl-D-glucoside. *Saccharomyces cerevisiae* has two alpha-glucosidases: a maltase and a saccharase. Moreover, the alpha glucosidases have a transglucosidasic activity that intervenes in the synthesis of polysaccharides and therefore in adhesins. A certain number of preparations are available on the market under the names of maltase, glucoinvertase, and glucido-saccharase. One unit frees 1 mmol of D-glucose from a p-nitrophenol-alpha-D-glucoside per minute at pH 6.8 and at 37°C. Maltose can also serve as a substrate and in this case, the unit provides 2 mmol of D-glucose per minute at pH 6 and at 25°C.

E.C.3.2.1.21 Beta-D-glucosidase or beta-D-glucoside hydrolase Also known by the name of gentiobiase, cellobiase, and amygdalase, this enzyme hydrolyzes beta-D-glucosides, beta-D-galactosides, alpha-L-arabinosides and beta-D-xylosides. As a whole, it divides the nonreducer terminal residues and frees beta-D-glucose. The reaction is written as:

beta-D-glucoside + H_2O = an alcohol + D-glucose

There is a wide variety of sources for this enzyme, for example; *Trichoderma viride,* and *Saccharomyces cerevisiae.* Aryl glucosides are generally better substrates than acyl compounds. This hydrolase also acts on alpha-D-fucosides, oligo-saccharides of galactose, galactomannans, and even galactolipids. The product comes from cultures of *Aspergillus niger* or green coffee beans. One unit of the enzyme, of mycosic origin, hydrolyzes 1 µmol of *o*-nitrophenol alpha-D-galactoside in *o*-nitrophenol and D-galactose in 1 minute at pH 4 and at 25°C. That of green coffee acts better at pH 6.5 and 25°C on the same substrate.

E.C.3.2.1.23 Beta-galactosidase Widespread in the bacterial world, this beta-galactoside hydrolase (or lactase) hydrolyzes beta galactosides and alpha-L-arabinosides:

beta-D-galactoside + H_2O = an alcohol + D-galactose.

One unit attacks 1 mmol of o-nitrophenol-beta-D-galactoside at pH 7 and at 37°C freeing in 1 min o-nitrophenol and beta-galactose. The forms on the market come from bovine livers, *Aspergillus niger,* or from the jack bean, *Escherichia coli*; and from rats' livers, *Diplococcus pneumoniae*. These examples give an idea of the variety of sources.

E.C.3.2.1.24 Alpha mannosidase This is an alpha mannosidase hydrolase acting on the alpha-D-mannose residues of alpha-D-mannosides. Moreover, it attacks the following compounds: methyl, benzyl, p-nitrophenol-D-mannosides, O-alpha-D-mannopyranosyl-(1-2)-O-D-mannopyranose, O-alpha-D-mannopyranosyl-(1-3)-O-alpha-D-mannopyranosyl-(1-2)-O-alpha-D-mannopyranosyl-(1-2)-O-D-mannopyranose, and mannosyl rhamnose. The enzyme also frees mannose from ovalbumin, ovomucin, and oromucin. Here again, is the example of a wide spectrum of activity showing the value of this hydrolase in the performance of releases. *Pseudomonas* stuck in certain mucus should be mentioned here. The unit of the product from the jack bean attacks 1 mmol of p-nitrophenol-alpha-D-mannose at pH 4.5 and at 25°C per minute.

E.C.3.2.1.26 Sucrase or invertase This is an alpha-D-glucohydrolase, beta-fructofuranosidase, or beta-D-fructanoside fructohydrolase. One unit degrades 1 mmol of saccharose per minute at pH 4.5 and 55°C. The marketed product comes from *Candida utilis* or simply from baker's yeast. Nevertheless, it should be pointed out that this hydrolase is present in many bacteria, in *Neisseria crassa* for example. It is inhibited by ethanol.

E.C.3.2.1.31 Beta-glucuronidase This very worthwhile enzyme is a beta-D-glucuronide glucanohydrolase acting based on the following relation:

beta-D-glucuronide + H_2O = an alcohol + D-glucuronate.

One unit frees 1 mmol of phenolphtalein from phenolphtalein glucuronide in 1 h at 37°C for a pH of 6.5 to 7. There are a variety of sources: rat's kidney or liver, mollusks (abalone, limpets, snails). The products most often available are of bacterial origin, from bovine livers or from mollusks.

E.C.3.2.1.35 Hyaluronidase Known under the names of mucinase, diffusion factor, and hyaluranoglucosidase, in the international nomenclature this enzyme is designated as being a Hyaluronate 4-glucanohydrolase. Its very substantial role endows it with the property of dividing the 1,4-beta-glucosidic bonds uniting N-acetylglucosamine or N-acetylgalactosamine sulfate to the glucuronic acid in chondroitin, 4 and 6 sulfates, and dermatan. It also has a transglucosylation activity. There is a wide range of preparations available from bovine or ovine testicles, *Clostridium perfringens,* or bee venom (*Apis* *mellifera*). There is not an officially well-defined unit of the activity of this complex enzyme that frees N-acetylglucosamine from a hyaluronate. It plays a particularly interesting role in the release experiments.

E.C.3.2.1.41 Pullulanase or pullulane-6-glucanohydrolase This enzyme hydrolyzes amylopectin and more generally breaks the 1,6-alpha-D-glucosidic bonds of pullulanes, amylopectins, and glycogen. The commercial preparation comes from *Enterobacter aerogenes*. In terms of enzymes most suitable for the attack of substrates, it is interesting to focus certain techniques on the destruction of substrates. Chitinase, for example, is indicated for releasing bacteria attached on crustacean or insect carapaces or other substrates possibly composed of chitin. This substance is extremely widespread from bacteria to the higher animals and including fungi, plankton, arthropods, etc., as confirmed by the inventories drawn up by C. Jeuniaux in 1963[307] in a work entirely devoted to this subject.

Aside from lipases and chitinases, many other enzymes can be used for destroying certain substrates of known composition or for breaking adhesins and even bacterial walls. Proteases are available on the market, the most common of which are bromelain from pineapples, ficin from fig trees, and especially papain extracted from papayas (*Carica papaya*). These hydrolases all act on casein and are, for this reason, indicated for the release of germs attached on dairy products. Peptidases and trypsin also attack proteins in conditions that will be summarized. These enzymes belong to the group E.C.3.4. of the international nomenclature. They are described as follows.

E.C.3.4.21.4 Trypsin This alpha and beta protease is very active and hydrolyzes peptides, amides, and esters involving a bond with the carboxyl group of arginine or L-lysine. It acts at pH 7.5 and at 25°C.

E.C.3.4.22.2 Papain or papainase This enzyme has been used since ancient times for the treatment of oozing wounds or for improving the tenderness of meat. It offers the advantage of acting at pH 7 and at 37°C, therefore in the best physiological conditions. Moreover, it acts on esterases and thiolesterases. We have studied a few of its therapeutic properties since 1949 with R. Babin and have shown that it has a slight bacteriostatic activity on *S. aureus*. We also used its proteolytic properties for liquifying pus, eliminating necrosed tissues, and facilitating healing (J. Brisou, A. Rigaud, J. Coste, 1952).[296] As a result, it acts directly on the substrates and frees bacteria, making them more sensitive to antibiotics. During past studies, we established that this sensitivity could be increased by 50%, notably in the presence of Streptomycine (J. Brisou, P. Babin, and R. Babin, 1974).[297]

E.C.3.4.23.1 Pepsin This hydrolase is active at pH 2 and at 37°C, but is obviously not suitable for all bacteria as the majority of them cannot withstand such acidity *in vitro*. There are different types A, B, and C, coded respectively: 3.4.23.1, 3.4.23.2, 3.4.23.3

The first type, A, acts more readily on phenylalanine and leucine bonds. Type B is less specific. It hydrolyzes gelatin and very discreetly hemoglobin, as opposed to type C which is very active on this hematoprotein.

E.C.3.4.24.3 Collagenase Collagenase or Clostridiopeptidase is only elaborated by an anaerobic bacteria *Clostridium histolyticum*, as we demonstrated in 1953.[298] It should not be confused with procollagenases which are much wider spread. The real collagenase acts on native collagen which is not the case for "procollagenases". Collagenase is particularly indicated for destroying tendons, aponeuroses capable of sheltering bacteria which would not be detected by the usual analysis techniques. It is active at 37°C and at a pH close to 7.4. These hydrolases act on dead tissues, cellular debris, pus, and certain cellulose- or chitin-based parietal structures, thus freeing a large number of attached or buried germs that are not detected by the analyses as previously specified.

This inventory, which is not exhaustive, is nevertheless sufficient for enabling the most common experiments and the applications of which will be given in a few examples. The following chart summarizes the optimal conditions of activity for all these hydrolases and serves as a guide for establishing experimental protocols.

4. OPTIMAL CONDITIONS OF USEABLE ENZYMES

This table summarizes the optimal conditions of the activities corresponding to the enzymes used. It will serve as a guide for specific release tests. For example, sialidase, which acts on numerous structures of membrane coverings, alone deserves being tested as extensively as possible. It interests hematologists and immunologists considerably. Its very effective action should not leave microbiologists indifferent.

EC	Enzymes	pH	T (°C)
3.1.1.3	Lipase	7.7	37
3.2.1.1	Alpha-amylase	6.9	20
3.2.1.2	Beta-amylase	4.8	20
3.2.1.3	Amyloglucosidase	4.5	55
3.2.1.4	Cellulase	5	37
3.2.1.11	Dextranase	6	37
3.2.1.14	Chitinase	6	25
3.2.1.15	Pectinase	4	25
3.2.1.17	Lysozyme	6.2	25
3.2.1.18	Sialidase	5	37
3.2.1.20	Alpha-D-glucosidase	6	25-37
3.2.1.21	Beta-D-glucosidase	5	37
3.2.1.22	Alpha-D-galactosidase	4	25

3.2.1.24	Alpha-mannosidase	6	25
3.2.1.35	Hyaluronidase	7	37
3.2.1.41	Pullulanase	5	25
3.4.21.1	Chymotrypsin	7.8	25
3.4.21.4	Trypsin	7.6	25
3.4.21.7	Fibrinolysine	7.1	37
3.4.21.11	Elastase	8	37
3.4.22.2	Papain	7	37
3.4.22.4	Bromelain	4.5	45
3.4.23.1	Pepsin A	2	37
3.4.24.3	Collagenase	7.4	37
3.4.4.12	Ficin	7	37
3.5.1.1	L-asparaginase	8.6	37
3.1.6.4	Chondroitin-sulfatase	7	37
4.2.2.4	ABC Eliminase	7.3	37
4.2.2.5	AC Lyase	7.3	37

Note: The first of these chondroitinases hydrolyzes the 6-sulfate groups of the units: 2-deoxy-D-galactose 6-sulfate of chondroitin sulfate; hence the synonym, Chondroitin-sulfate-sulfohydrolase. Chondroitin ABC Lyase acts on chondroitin 4-sulfate, chondroitin 6-sulfate, dermatan sulfate and modestly on hyaluronate. Elimination of residues delta-4, 5-D-glucuronates of polysaccharides, containing the bonds 1,4-beta-D-hexosamyl, and 1,3-beta-D-glucuronosyl or 1,3-alpha-L-iduronosyl. Chondroitin AC Lyase eliminates the residues: delta-4,5-D-glucuronates of polysaccharides, containing 1,4-beta-D-hexosamyl and 1,3-beta-D-glucuronosyl bonds.

2 PRACTICAL APPLICATIONS

N.B.: Culture media — The composition of media used in this work is given in the Handbook of Microbial Media by Ronald M. Atlas, published by CRC Press in 1993.

The Coccosel medium is pratically identical to the EnteroCoccosel-Agar (p.338) and to the Enterococcus-Agar (p.339). This medium is also used for isolation of Listeria monocytogenes.

1. TECHNIQUES AND METHODS

The samples to be treated: sand, sediments, soils, shell grindings, a variety of foodstuffs, etc., must first be perfectly homogenized. They are then submitted to the action of one or more enzymes depending on the circumstances and the programs planned. For routine operations, a volume of 10 ml of a sample is enough. This volume, placed in a single-use sterile tube, receives the enzyme solution according to the modalities to be specified below.

1.1. Quantitation of the Enzymes

The quantity of enzymes to be put into action constituted the first problem posed by this methodology. Attention should indeed be paid to the macromolecular protein nature of the enzymes since such substances could justifiably be considered as nutrients capable of promoting microbial proliferation. The experimental route made it possible to answer this first interrogation without any trouble. It is indeed a known fact that enzyme proteins are mediocre substrates. It was nevertheless essential to specify the limits of this "stubborn" character.

Various common commercial preparations: hyaluronidase, cellulase, and pectinase solubilized in water buffered at pH 7 and sterilized by filtering through a Millipore filter, were seeded with fresh strains of *Staphylococcus aureus*, *Pseudomonas aeruginosa*, *Escherichia coli*, *Streptococcus faecalis* and *Candida zeylanoides*. There was no visible growth after 48 h of incubation at 37°C for bacteria, or at 28°C for yeast in the presence of quantities of

153

enzymes less than or equal to 1 mg/ml. This value served as a reference for the entire experiment. On this basis, stock solutions were prepared at 1% in distilled water. Sterilized by filtration (filtering syringes are now available enabling very rapid operations), these solutions maintained their activity for over 8 days at 4°C. It is easy to package small ready-to-use volumes and keep them in the freezer.

We made sure that the filtration did not harm the activity of the preparations and that, in certain cases, it was even high, with the preparations filtered only with the controls (for example, amylases and cellulases). At the time of use, simply add 1 ml of the stock solution to 10 ml of the sample to be treated. The enzyme level does not go over the mg/ml. It goes without saying that this quantity can be easily reduced by taking into account the activity of the commercial preparations that, for a single hydrolase, can range from 100 units to 30,000 U/mg. This will therefore be mentioned in the experimental protocols. The concentration of the stock solutions is, for convenience, maintained at 1%. The active quantities distributed are adjusted by the experimenter depending on the theoretical units indicated on the product packaging. The volume distributed therefore ranges from 0.3 to 1 ml depending on the circumstances.

1.2. Conservation of the Enzyme Solutions

Solutions Kept 8 Days at + 4°C: Hyaluronidase or Pectin 1 mg/ml

Media	Sediment I			Sediment II		
	Hyal	Pect	Control	Hyal	Pect	Control
Mueller Hinton	1160	+++	80	480	+++	220
SS.	450	+++	0	0	0	0
Endo	860	+++	45	400	+++	25
Brilliant green	+++	+++	+++	+++	+++	160
Coccosel	2	16	1	2	2	3

Note: +++ = invasive culture, with impossible enumeration. It was impossible to conclude that the solutions kept 8 days at 4°C had a proper action, at least for the 2 enzymes considered.

1.3. Standard Method

The mixtures enriched with enzymes are incubated at 37°C for 2–24 h in darkness and constant agitation; 1 to 2 h are sometimes enough for ensuring satisfactory release. Control samples, without enzymes, are placed in the same conditions of temperature and agitation.

After incubation, the mixtures are seeded on appropriate culture media, either standard or specific, enriching, selective, differential, etc. The choice is unlimited. The incubation temperatures for these media are also chosen depending on the germs sought. For example, the choice is between 18–20, 28,

37, 44, and 55°C. The experimenter should seed very small volumes after the action of the enzymes to avoid any invasion of the cultivable surfaces.

The experiment led us progressively to limiting the inoculums to volumes of 25–50 µl for a 10-cm Petri dish. Once the colonies were enumerated, it was easy to choose among the most characteristic, using the conventional methods to identify the germs that gave birth to them.

1.4. Specific Method

The analysis of the overall chart concerning the optimal conditions of activity of the enzymes suggests the application of this data in the pursuit of certain programs. It goes without saying that if acid pH of 4 or 5, or temperatures close to 50°C are not very favorable for the growth of many bacteria, such conditions are, on the other hand, perfectly suitable for others, for which they facilitate both the release and growth. As a result, germs can be evidenced that would have normally not been detected by usual investigations.

In this way, we isolated numerous *Bacillus*, uncommon *streptococci*, micrococci, and other bacteria after action of enzymes in conditions of acid pH and temperatures of 40°C and even 50°C. This was the case, for example, with hyaluronidase at pH 4.5 and incubation at 55°C, with cellulase at pH 5 incubated at 37°C. The choice is practically unlimited and the results often surprising!

Such very selective methods lend themselves to multiple applications, notably in the field of the foodstuffs industry and dairy industry which are especially interested in thermophilic bacteria withstanding high acidities. Microbiology of soils, marine and lake sediments, as well as oil and mineral deposits, are other fields of possible applications.

1.5. Stages of Release

Figure 50 illustrates the stages of release.

2. EXPERIMENTS WITH ONLY ENZYMES

2.1. On Marine Sediment

Marine sediment from the Gulf of Giens (depth: 8 meters), treated using pectinase 100 units/10 ml of sedimentary solution, with an incubation at 28°C pH 6.5 for 1 h only, in constant agitation. The cultures of soy trypticase agar are at 37°C. The number of colonies developed per gram of dried sediment.The volume of the inoculum is 0.2 ml.

Control	Pectinase	Difference	%
362	650	288	44.0
373	687	314	45.7
717	1145	428	37.5

Homogenous suspension of sand, sediment, foodstuffs, tissular grindings, etc.
Liquid : water, milk.
Stock solution of enzymes sterilized at 1% concentration.

1st stage - enzyme solution :

Volume: 10 ml
Enzyme solution of 0.5 ml to 1 ml
Incubation: from 1 to 6 hrs, or in outstanding circumstances, longer.
Continuous agitation at 37°C and pH 7
The variants are on :
The incubation temperatures and the pH
The possible addition of one or more antibiotics (50ug/ml).

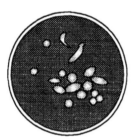

2nd stage - cultures :

Standard or selective solid media
Enumeration of the colonies
Isolation of the most characteristic - Identifications.
For these enumerations, the experimenter should dilute the supernatant of very rich mixtures. These dilutions then lend themselves quite well to filtrations through Millipore filters and to seedings on all the possible differential media.

Figure 50. Sample stages of release.

Following is the same experiment with hyaluronidase at 5000 units/20 ml with contact under constant agitation 1 h at 28°C. The inoculum is 0.2 ml and the number of colonies is per gram of dry sediment

Control	Hyaluronidase	Difference	%
261	572	311	54.0

Another experiment, after 24 h of enzymatic action yields:

Control	Enzyme	Difference	%
68	283	215	76

2.1.1. Action of Hyaluronidase on Marine Sediments

The amount of enzyme is 110 units for 10 ml of suspended sediment with a contact of 1 h at 37°C under constant agitation. The results here are expressed in number of colonies/g of dry residue.

Control	Enzyme	Difference
58	86	32.5%

And after 24 hours of contact on two different sediments:

78	1.173	93%
411	13.853	97%

Extension of the contact apparently promotes the multiplication of germs thanks to oligosaccharides, monosaccharides, and glucosamine released by the enzyme. The germs released as from the first hours proliferate. We showed this in the presence of antibiotics in cooperation with B. Makhlouf.

Prolonged action of the enzyme over a 24 h period with the number of colonies developed per 1 g of dry sediment:

Media	Control	Enzyme
Coccosel	0	400
Hektoen	0	450
Endo	0	370

In official data, the conventional method provides 0 colonies!

After release with alpha-amylase: Coliforms = 750,000/100 ml and *S. faecalis* = 750,000/100 ml.

This experiment confirms that enzymatic actions limited to 1.5 h at 28°C are sufficient for ensuring a release with surprising results. Such is the case, for example, with hyaluronidase, which on Endo media and soy trypticase agar, enabled enumerations of colonies considerably higher than the controls, namely: 81,000 and 12,000 as opposed to 2300 and 4300!

2.1.2. *Glucuronidase Experiment*

	AK	AL
Endo	++++ Pink colonies	++ 2 types of white and pink colonies
Brilliant green	++++ Blue colonies	Rare, two light green types
Coccosel	Four black colonies (*S. faecalis*)	20 Small light colonies
SS.	Seeding of fine colonies, + 16 thick, dark pink	Small pinkish colonies

Note: 1 mg/ml. 18 h period of action at 28°C and pH 7 for sediment AK and pH 5 for AL Cultures on agar with starch. AK = invasion by 4 varieties of colonies and AL = invasion by *Bacillus.* Cultures on other media, incubated at 37°C.

2.1.3. Cultures on Differential Media

UFC Counted on the Dishes — Inoculums = 0.20 ml

Media	Control	Cellulase
Endo	E. coli 12	E. coli ++++
Hektoen	280 L+ and L–	Carpet of L+
Brilliant green	305 L+ and L–	Carpet of L+
Blood agar	580 Hemol.	++++ Hemol + and –
Coccosel	8	++++ black
Sabouraud	76	129
Fungiphil	23	150

Note: Inoculums, 0.20 ml. Release by several enzymes. Cultures on differential media. Experiment on beach sands.

Data from Makhlouf, B., Brisou, J., and Stevinino, J. C.I.E.S.M., 1983.

UFC Counted on the Dishes — Inoculums = 50 µl

Media	Control	B-amylase	Glucuronidase	Cellulase
Hektoen	10 L–	++++ citrobacter	++++ E. Coli	0
Endo	0	800 L+		
Hektoen	0	37 L+ and L–		
Coccosel	0	450		

Note: L = acidification of lactose (+) or not (–). Hemol = hemolysis

In the last example, the expression of the results according to usual standards yields Coliform: 800,000 and 740,000/100ml and S. faecalis: 800,000/100 ml.

2.2. Experimentation on Mussel Grindings

The enzyme action time is 1.5 h at 28°C with inoculums at 0.20 ml. We maintained incubation at 28°C to avoid pullulation of numerous microbes present in such a product, despite an intentionally very short time. We then extended the experiment with a 24-h incubation.

Enzyme action time	1.5 h	1.5 h	1.5 h	24 h	24 h	24 h
Nutrient	Agar	Endo	TCBS	Agar	Endo	TCBS
Control	2300	4300	666	50,300	10,000	330
Hyaluronidase	81,600	12,800	866	100,000	800	33,300
Pectinase	8000	660	200	56,000	1000	0

Note: In other experiments with the same technique, with incubation at 37°C and inoculums of 25 µl, the results on TCBS medium were: Control – 1533, Hyaluronidas – 7350, and Pectinase – 17,300 per gram of dry residue.

The action of pectinase in the release of *S. faecalis*, and the considerable variations of the results depending on the seeded media is particularly noteworthy and very typical. Scanning electron microscopy controls showed the reality of releases (J. Brisou, L. Lecarpentier, and B. Makhlouf, 1982):[300] *The fragments not treated by enzymes are very abundant in bacteria, whereas the treated fragments are particulally unoccupied.*

2.3. Treatment of Infected Urine

Urine treated by a mixture of hyaluronidase and pectinase at 1000 U of the former for 5 mg of the latter. The volume of urine is 10 ml and the incubation is 3 h at 37°C for 0.20 ml inoculum.

Number of colonies per 10-cm diameter dish

Media	Control	Enzyme
Coccosel	5	10
Blood agar	604	+++ Invasion
Endo	+++	+++ Uncountable

Note: Identification: *S. faecalis* and *E. coli*.

In this experiment, there were no qualitative differences but there were very wide quantitative deviations, notably on the blood agar and the Endo media. The inoculums were too abundant. Following such results, we limited them to 25–50 ml.

2.4. Experiment on Inert Powders

Experiment on *S. aureus* (Institut Pasteur strain)
1st stage: Attachment — results in UFC

Control	Montmorillonite	Kaolin	PVP
++++	0	0	2

1st stage: enzymatic release of the attached germ (UFC)

Enzymes			
Alpha-amylase	0	10	30
Beta-amylase	22	5	5
Amyloglucosidase	6	0	0
Fructosidase	0	3	4
Beta-glucuronidase	5	1	0
Hyaluronidase	0	67	13

This experiment highlights the importance of the substrate for one and the same germ, and it shows the obvious release of montmorillonite in the presence of beta-amylase on kaolin, using hyaluronidase, and on PVP with alpha-amylase.

2.5. Association of Enzymes-Antibiotics

Experimental system
Single-use sterile tubes with 6.5 cm bottoms
Suspended material: 10 ml
Enzyme solutions: 5 ml
Antibiotics: 50 mg/ml
Sterile distilled water: 5 ml
Total volume: 20 ml

The mixtures were incubated at 37°C for 24 h under constant agitation. $50 \mu l$ seeded per 10-cm Petri dish lined with the appropriate culture media. Glucose quantitations performed at regular intervals show a rapid release of glucose, confirming the obvious intervention of hydrolase. The commercial, impure pectinases also release glucose in addition to galacturonic acid. The activity of the enzymes acts on both the adhesins and the natural substrates. The presence of antibiotics slows the consumption of sugar. It starts over again after a few hours thanks to the microorganisms resistant to antibiotics. Microscopic tests show the wealth of mixtures in dilacerated vegetal debris, digested by the enzymes.

Association of enzymes and antibiotics on beach sands. Number of colonies/ml counted on the dishes (suspended material) — Inoculums: 50 μl

Media	Hektoen	Endo	Coccosel
Control	65	0	0
E1	5600	16,000	3000
E1 + A1	0	0	0
A1	0	0	0
Control	30	2400	0
E2	6000	12,000	12,000
E2 + A1	0	60	0
E3	7000	13,000	12,000
E3 + A1	10	14,000	4000
Control	280	1200	400
E4	5600	80,000	60,000
E4 + A2	5600	701,000	44,000

Note: E1 = cellulase, E2 = amyloglucosidase, E3 = pectinase, E4 = alpha-amylase, A1 = thiophenicol (50 mg/ml), A2 = bacitracine (10 mg/ml).

Data from Makhlouf, B. and Brisou, J. *C.R. Acad. Sci. Paris,* 295, 679, 1982.

The principle was to mix the marine sediments with the enzyme solutions in the presence of antibiotics and to perform, after a certain period of contact, bacterial enumerations to compare the results with those obtained using nontreated sediments. The number of sediments is 12, from the Toulon shoreline.

Enzymes:
  Alpha-amylase (E.C.3.2.1.1) from pig pancreas
  Amyloglucosidase (E.C.3.2.1.3) from *Rhizodus*
  Cellulase (E.C.3.2.1.4) from *Aspergillus niger*
  Pectinase (E.C.3.2.1.15) from *Aspergillus niger*
  Stock solutions at 10 mg/ml in distilled water
  Sterilization by filtration through 45 µ-Millipore filters

We made sure that this filtration did not cause any decrease in the enzymatic activity. The treated solutions are often even more effective than the untreated solutions. Conservation of solutions occurred at 4°C.

Antibiotics:
  Bacitracine at the final concentration of 50 mcg/ml.
  Kanamycine, polymyxine, and thiophenicol in the same quantities.
Culture media:
  For the enterobacteria: Endo media and Hektoen media.
  For *S. faecalis* and Gram-positive germs: Coccosel.
Quantitations of glucose in certain mixtures:
  Technique with glucose oxydase.

Results

Glucose quantitations —The quantitations performed at regular intervals show a rapid release of hexose (2 h) under the enzyme action, confirming their activity. We know that pectinase releases galacturonic acid but it is specified by the producer that the preparation also contains other enzymes (its purity is not guaranteed) and this comment naturally explains the appearance of glucose in the mixtures.

The activity is both on the natural substrates and on the attachment apparatuses of the germs. In the presence of antibiotics, consumption of sugar is slowed down. It seems to be the fact of resistant and selected germs. Direct microscopic examinations show numerous vegetal debris partially digested by the enzymes, notably cellulase and pectinase. A release selecting Gram-positive bacteria can be witnessed, notably as far as the germs capable of developing on the Coccosel media are concerned. The apparently most valuable enzymes belong to the group of amylases, glucuronidase, with which the values recorded reach the maximum levels: 80,000 to 100,000 colonies per milliliter for inoculums of 50 µl, even in the presence of antibiotics (Bacitracine, for example).

Cultures — The figures shown in the previous chart confirm and specify what was observed in 1979 and 1980. The number of colonies developed on the different media reach the maximum values with the mixtures treated with enzymes or by the mixtures of enzymes and antibiotics. Certain cases are particularly interesting and significant.

The differences between the mixtures, with or without antibiotics, are often limited or nonexistent (EA + A2, for example) since the released germs are resistant to the antagonists. It is noteworthy that the untreated sediments provide modest cultures which make it possible to consider them in compliance with health standards. Yet, the complete opposite is true after the enzymatic treatments: the figures reach values far over the standards. Numerous colonies were isolated for identification. We can already point out the value of an authentic *E. coli* strain released in the mixture of amyloglucosidase and glucuronidase in the presence of thiophenicol and bacitracine in the quantities of 100 and 150 mg/ml, respectively. This strain proved to be resistant to all the common antibiotics. The frequency of *Bacillus* should also be noted since it suggests the development of an isolation method of the representatives of this bacterial group by an association of judiciously chosen enzymes and antagonists.

Comments — These results lead to some interesting concepts; in all likelihood, polysaccharases act on two levels, on the natural substrates of bacteria and on adherence polysaccharides that the latter synthesize.

2.6. Experiment Using Bacteria Marked With Radioelements

Radioelements have been used for a long time in biology and they make it possible to monitor the evolution of bacteria in a wide range of circumstances. The active radioelement chosen, 14C, 32P, 3H, for example, is introduced in the most favorable culture media for the growth of germs. After 24 h of growth, the latter are harvested, washed, and put back into suspension in a physiological solution that will be used to contaminate the various media, where their behavior will be monitored by scintigraphic measurements. The graph figures below express the number of "strokes" per minute. The curves correspond to the evolution of the radioelement of the medium over time. These measurements were completed by autoradiographies when the experiment subjects were represented by mollusks.

2.6.1. *Attachment of Labeled Bacteria on Shellfish*

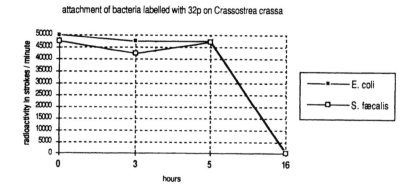

attachment of bacteria labelled with 32p on Crassostrea crassa

An experiment similar to the previous one, but using a slower marker, was performed on two mollusks: *Venus verrucosa* and *Glycymeris glycymeris*. Anin is one of the main amino acids of the cell wall structure in Gram-positive bacteria. It makes a choice marker. Bacteria marked following a culture in a rich media at 37°C for 24 h were, after washing, put into two aquariums filled with natural seawater. One served as a control, the other hosted the two bivalve mollusks.

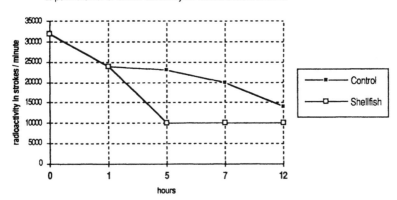

Experiment with S. fæcalis labelled by 3H-alanine and two shellfish

The curves presented here indicate drops of radioactivity in the water of both aquariums. This drop in the control corresponds to the adherence of germs on the glass walls. The second, much more noticeable, corresponds to the adherence on the shells. It was confirmed by autoradiographies of fine curves of each of the mollusks. We can confirm, as from the first hour, that the bacteria began to invade the shells on the branchia, and that after the third hour, the mollusk was entirely colonized, notably on the hepathopancreas.

2.6.2. *Evolution of a Microbiocenose in a Closed Space*

Following the experiments performed with bacteria marked either using 32P or with 3H-alanine, we were led to monitor the variations of the radioactivity of natural seawater contained in aquariums. Some sheltered bivalve mollusks; others, without shells, were used as controls. Analysis of the curves show a rapid drop of this radioactivity not only in the presence of the shells, but also, a weaker drop in the control aquariums. This drop corresponds to the attachment of bacteria (*S. faecalis* and *E. coli*) on the walls of the recipients. Extension of the observations made it possible to record a rise in the radioactivity of the water, which in the control aquariums obviously testified to a release of part of the germs attached during the first stage of the experiment.

We previously mentioned the Dutch work published by Driessen et al.[272] in 1984, devoted to the attachment of *Streptococcus thermophilus* on sheets of stainless steel submerged at the same time in pastuerized milks and in raw milks. The adherence of thermophilic germs is practically the same in both types of milks during the first 4 h; beyond this time period, the authors noted a higher attachment rate in the pasteurized milks, contrasting with a drop in the raw milks.

These two experiments can be compared. Driessen and his colleagues think that pasteurization destroys factors opposed to attachment. We can more simply suggest that in both cases — raw milk or natural seawater — the autochthonous microbiocenose begins to proliferate between the 4th and 8th hour, depending on the conditions of temperature and the concentration in foods. A variety of bacteria, yeasts and lower fungi develop polysaccharases which quite naturally cause a release of part of the attached germs.

In 1946, we monitored the variations of bacterial populations in natural seawater stored according to ZoBell's techniques.[243] Substantial fluctuations, which were poorly explained at the time, were observed, but in light of what we know now about the mechanisms of adherence and release, these variations have become easier to understand. The successions of adherence and release are governed by the contents in hydrolases of all kinds, elaborated by the microbiocenoses present in the microenvironment considered (glass or plastic jar, tank, etc.). It should be noted that the results of analysis of water or any other liquid, performed several hours after sampling are, from the start, marred by errors.

As from the first hour, it can be estimated that approximately 1/10th of the microbiocenose is attached to the walls of the recipient, thus escaping any attempt at enumeration. This aspect of natural release is far from being insignificant in practice whether an artificial or natural microenvironment is concerned. Numerous applications of this experimentally confirmed concept can be foreseen, such as the monitoring of swimming pool health standards. We noted that the analysis of the walls provide quantitative and qualitative results totally different from those recorded for the pool water.

2.7. Special Case of Fat

Enzymology naturally advises the use of lipases or, if need be, pancreatin, which is rich in lipase. In the investigation of the causes of an epidemic of typhoid fever (*Salmonella typhi*) occuring in the Port of Brest (France) in 1934–35, butter was finally suspected.[302,303]

The methods in use at the time only providing negative or excessively conflicting results, I fell back on the emulsionating power of bile, which had a dispersion effect of the substrate far ahead of time!

Thanks to this process, it was easy for me to isolate a certain number of salmonella strains and to assert that they were responsible for the epidemic. This was also the opportunity for determining the survival of this enterobacteria in butters depending on their age and their acidity; the fresh butters proved to be the most dangerous. The technique suggested (J. Brisou, 1935) is still valid and among the easiest:

Carefully sample 5 g of butter or any other fat, and place it in a small sterile tube containing 15 ml of sterile (cow or pig) bile. Place in a 37°C double boiler, in constant agitation, for 10 to 15 min. Shake vigorously to perfect the emulsion that will be seeded on the appropriate media depending on the research to be undertaken. It goes without saying that this method will only be applied for evidencing germs insensitive to bile: enterobacteria or *Listeria monocytogenes*, for example, the growth of which is even promoted by this natural secretion.

2.8. Experiment by Richelle Maurer

E. Richelle Maurer[304,305] for example, specifically studied well water in Kisangani, Zaire in 1984–85. She used three enzymes: alpha amylase, cellulase, and hyaluronidase. Three media were seeded after the action of the hydrolases: Trypticase-glucose-soy, medium with eosine and blue methylene, and MacConkey medium. The sometimes considerable differences were recorded between the controls and the water treated with the enzymes. The highest values were obtained after the action of alpha amylase and hyaluronidase. The growth rates could reach 10,000 and 15,000 in relation to the controls. It was possible to isolate certain treated samples: two strains of *Salmonella* after hyaluronidase action, three strains of *S. faecalis* after realease by alpha amylase, several strains of *Serratia marcescens* of *Klebsiella*, *Enterobacter*, and more rarely *Acinetobacter* which, using the usual methods, were not revealed.

2.9. Experiment by E. Ebiou

In 1988, E. Ebiou,[306] a colleague of R. Moreau, devoted a major study to the mycobacteria of soils, litters, muds and vegetals of an alpine region. During his research, the author implemented a few enzymatic releases, including the essential one that follows: The analysis of 20 samples treated by the conventional methods did not enable any isolations of mycobacteria. However, after treatment by enzymes, he obtained seven positive results. The hydrolases experimented were amyloglucosidase, cellulase, and hyaluronidase. Three strains were released by cellulase, hyaluronidase, and alpha amylase; another was released by cellulase alone, and another under the action of hyaluronidase; and two strains were released by hyaluronidase and cellulase. The enzymes were not associated in any of the cases; they acted separately. After action of hydrolases, the mixtures were decontaminated by cetylpiridinium chloride at 0.75%.

Following these results, the author considered that the mycobacterial population of the studied environment was underestimated, for lack of an appropriate technique making it possible to detect the bacteria. During his investigation, he obtained 35% of positive cultures after enzymatic release, whereas the conventional methods did not reveal any mycobacteria. This study, although limited, is nevertheless noteworthy.

2.10. Experiments by Patrick Brisou

Bacterial Release by Specific Atack of Substrate

Patrick Brisou. Laboratoire de l'Hôpital d'Instruction des Armées, Toulon (France), February–May 1994.

Action of Cellulase on Vegetal Substrates

1. *Material* — Experiments were done on leaves of identical size from a lilas *Syringa vulgaris.*
2. *The enzyme* — The enzyme was a SIGMA cellulase, from *Penicillium funiculosum.* Activity: 10 units/mg — Freeze-dried and stored at –20°C. The enzyme solution is prepared by dissolving 100 g of this powder in a liter of sterile physiological solution, containing 0.9% NaCl. We thus had a solution titering 1000 units/ml. The pH is adjusted to 7.0.
3. *Culture Media* — The four culture media used are marketed, ready-to-use, solid media: Chocolate agar, Mueller-Hinton agar, Sabouraud's agar with chloramphenicol added, blood agar with (NAC) nalidixic acid and colistin added.

Methodology

Preparation of the samples — Each leaf was cut in two lengthwise using a sterile scalpel on an operating field. A test half and a control half were obtained from each leaf. The breakdown in couples was performed randomly and blindly. The bottles were then filled with either 5 ml of cellulase solution or physiological solution, then shaken and incubated in a double boiler at 37°C for 3 h.

Enumeration of microbial agents — After incubation, agitation of each bottle was performed before seeding 100 µl on each medium using the rake technique. The media are then incubated in the furnace: the chocolate agar at 37°C, the ANC blood agar at 37°C in anaerobiosis, and the Mueller-Hinton and Sabouraud's agars at 28°C in aerobiosis. Readings were made every 24 h for 5 days using a counter and a magnifying glass on a squared screen. All the small or large colonies were enumerated in this way. Identification was only performed on the dominant colonies and is limited to the diagnosis of families.

Results and Comments

Quantitative Results — Fifty samples were thus quantified using four different media. The results of the enumeration of the bacteria and yeasts are expressed in units forming colonies (UFC) per ml of solution.

Chart of cellulase experiment:

Mean of 10 samples in series	MH		ChA		NAC blood		S	
	Enzymes	Control	Enzymes	Control	Enzymes	Control	Enzymes	Control
Series I	192	99	48	41	5	6	70	66
Series II	98	58	63	28	6	4	35	31
Series III	163	84	47	39	3	2	48	42
Series IV	200	101	70	51	13	10	64	17
Series V	149	83	49	41	6	5	51	44

Note: Culture media: MH = Mueller-Hinton agar, ChA = Chocolate agar, NAC blood = Blood agar + Nalidixic Acid + Colistin, S = Sabouraud's agar.

The statistic analysis of these results was possible thanks to the methodology followed: large number of samples, random, blind breakdown, pairing of tests in couples. The method used for comparing the cellulase technique of the control is the couples method. The analysis shows that both methods differ significantly with a risk below 1% and this is the case for all the media with the exception of the NAC blood agar. The bacterial release technique using cellulase when the media is capable of containing cellulose provides results significantly higher than those of the control.

For the isolation of the bacteria, the Mueller-Hinton medium provided the highest performance irrespective of the method used. The significant difference that is observed with the chocolate agar may appear paradoxal, since the latter medium is much richer and enables the isolation of most of the demanding bacteria. The difference is in all likelihood due to the respective temperatures of 37 and 28°C. The latter certainly promotes the growth of germs in the environment.

The isolation of yeasts on the Sabouraud-chloramphenicol medium was much higher than that obtained on the other agars. To avoid weighing down the protocol, we did not multiply the variety of the culture medium. However, it is obvious, as shown by the results obtained with the Sabouraud's medium, that selective media using a heterogenous surface such as a leaf as a point of departure, can increase the number and the variety of the bacteria recognized.

Action of Chitinase

Chitin is a polysaccharide very widespread in nature, notably in insects and crustaceans; microorganisms adhere to the surface of these cuticles, where they attach themselves, thanks to their adhesins, on this especially favorable substrate. Within the framework of microbial release, it is worthwhile to evaluate the action of chitinase. Certain marine bacteria, such as *Vibrio alginolyticus* and numerous others, do indeed have this enzyme (J. Brisou et al., 1969).[294] A few of these bacteria are responsible for a deadly disease in lobsters, by destruction of the exoskeleton (*V. alginolyticus, Vibrio parahaemolyticus, Aeromonas,* etc.).

Material — Gray shrimp, *Grangon grangon,* of Mediterranean origin, were chosen for this study. They were purchased alive.

Chitinase — SIGMA chitinase comes from *Streptomyces griseus.* It is in 12 mmg bottles. It titers 860 U/g.

Test solution — 36 mg of enzymes in 30 ml of sterile physiological solution with a pH adjusted at 6.5. The solution contains in fine 1 U/ml. The reading of the cultures was performed every 24 h for 5 days. The identifications were reserved for dominant colonies and limited to the diagnosis of families or genera as in the previous study.

Experimentation — Each shrimp was shelled using a sterile scalpel and tweezers on an operating field. The external cuticle was extracted, avoiding any contamination by the intestinal contents of the crustacean. The cuticle was quickly broken and weighed on a precision scale. 20 mg samples were divided up in couples randomly and blindly. Sterile bottles then received 1 ml either of chitinase solution or of physiological solution, then were shaken and incubated in a double boiler at 30°C for 12 h.

Microbial enumerations — After incubation, each of the bottles was shaken before seeding 100 µl of this preparation on each of the solid media using the rake technique. The cultures were incubated at 30°C in aerobiosis. As in the previous experiment, readings were made every 24 h for 5 days. Identifications on dominant colonies. Diagnosis was limited to families and generations.

Results and Comments

Thirty samples were analyzed using three different media. Enumeration of bacteria and yeasts was expressed in units forming colonies (UFC) per ml of preparation. The Mueller-Hinton medium is the one with the highest performance for the isolation of bacteria. TCBS agar is, on the other hand, selective for *Vibrionaceae.* On both media, the number of colonies was significantly higher with the technique using chitinase with a risk of 0.1%. The number of yeasts isolated was too low to allow a comparison of the two enumeration techniques.

Chart of Chitinase Experiment

Mean of 10 samples in series	MH		TCBS		S	
	Enzymes	Control	Enzymes	Control	Enzymes	Control
Series I	242	90	63	31	6	3
Series II	278	105	63	23	23	9
Series III	282	87	51	16	7	2

Note: Culture media: MH = Mueller-Hinton agar, S = Sabouraud's agar, TCBS = Thiosulfate-citrate-bile-saccharose (specific media of vibrionaceae).

Qualitative Results — The bacterial families isolated were relatively limited. The most frequent genera were: *Enterobacteriaceae, Pseudomonadaceae, and Vibrionaceae*. Chitinase made it possible to evidence a larger number of vibrions, without a larger variety of species. We noted the absence of *V. alginolyticus*, normally rich in chitinase. This detail could have distorted the results. The isolation of yeasts was not significantly different in both techniques.

Conclusion for These Experiments

The performance of chitinase in the release of microorganisms, on the cuticles of crustaceans, and more precisely on Mediterranean gray shrimp, is notably higher than in the control samples. It would be worthwhile to compare this technique with those that call on mechanical means: for example, strong agitation or ultrasounds which are always traumatizing and not especially specific. Other experiments could be planned, making it possible, among other things, to evaluate the potential action of chitinolytic bacteria, *Vibrio alginolyticus*, on certain crustaceans from temperate and warm seas.

POSTFACE

POSTFACE

The enzymatic release of microorganisms from their natural shelters now appears to be a well-established fact, insofar as credit is granted to the experimental results recorded since 1978, which are now in part confirmed by other circumstances.

This release clearly acts on various levels:

- Directly on microbial adhesins
- On the structures of the media sensitive to enzymes
- On the magmas of polysaccharides organized by the germs themselves during the colonization of the interfaces, whether inert or live
- On aggregates

It is obvious that from the first hours of contact with hydrolases, the release is for the most part guaranteed. Generally, 2–4 h of incubation are sufficient, in constant agitation and at wisely chosen temperatures based on the programs to be undertaken. Some release experiments were limited to 1 or 1.5 h, as shown by some results; whereas, in other circumstances the enzymatic action was extended for 20 to 24 h.

The enzymes free oligosaccharides and monosaccharides favorable to the growth of the released microorganisms. Extending the contact between the sample and the hydrolases obviously leads to very high counts of colonies which will not come as a surprise if the development of the events are understood. The studies performed with mixtures of antibiotics and enzymes provide the demonstration of this. They confirm the liberation of glucose from the first hour. The levels rapidly reach 2g/l. The monosaccharide is consumed by the germs that resisted the antibiotic action. There is no doubt about this point. The intentional extension of hydrolysis is nevertheless, irrespective of what one may believe, of unquestionable interest. Emphasis should once again be placed on the fact that a released, free germ cultivates more easily than a trapped microbe caught in a gangue or buried in tissues.

Based on this principle, the specific liberation of certain microbes offers them unquestionable advantages over the other representatives of a microbiocenose and facilities of colonization which rapidly give them the lead over the other populations. Glucides such as glucose (easily assimilated), maltose, glucuronic acid, glucosamine, etc., represent some of the substrates favorable to growth. This is, for example, the case for salmonella, fecal streptococci, Enterobacter, Citrobacter, or other easily cultured bacteria. The results presented in the charts are convincing enough. It is easy, taking into account these experimental data and technical details shown, to vary:

- *The action time of the enzymes chosen* — This action can be limited to 1–2 h, or extended for an average duration of 6–8 h, or even 20–24 h.
- *The optimal conditions of hydrolases activity* — Therefore, at specific temperatures and pH. Such conditions are not favorable for all microorganisms, some are frankly even hostile. Particularly noteworthy selection means are available which can be a function of the specific resistances of bacterial groups to the chosen temperatures and pH.
- *The sometimes limited, sometimes widened specificity of the hydrolases* —Some act on a very limited group of substrates, whereas others, such as cellulase, amylases, chitinase, etc., have a wider spectrum.
- *The associations of enzymes and antibiotics* — They are limited, and make it possible to imagine a wide range of specific isolation media.
- *Separate enzymes* — It is indeed just as convenient to work with separate enzymes as with mixtures. We often used the association of Hyaluronidase + Pectin which provided interesting results. However, separate hydrolases enable better identification of the type of anchorage polysaccharides or adherence magmas. With the knowledge, for example, that many *Pseudomonas*, including *Pseudomonas aeruginosa*, adhere due to alginates (similar to laminarins) for their release, it is indicated to call on cellulase, or sialidases, in the absence of alginase which is (unfortunately) not yet available on the market. The results obtained may contribute to fighting adherence on certain tissues. The enzyme mixtures must in any case be limited; associations with proteases should notably be avoided.

Release, an essentially biological phenomenon, is obviously not always a success. The operator should expect a few failures. There is never a total release of germs. Experiments provide proof of this and make it possible to supply some reasons.

Success is related to the microorganisms themselves, to the type of their adhesins, but also to the media. By experimenting with germs marked with 32P (Enterobacteria, *Pseudomonas, Streptococcus faecalis,* yeasts) and inert powders such as montmorillonite, kaolin, polyvinyl-pyrrolidone, intentionally contaminated, we noted considerable differences in the releases with deviations from 17 to 80% depending on the specific properties of the adsorbents. The best results were obtained under the action of hyaluronidase, with *S. faecalis* attached on kaolin and polyvinyl-pyrrolidone. All these results are without a doubt surprising.

Beach sand, water, and mollusks of common contamination, considered perfectly in compliance with safety standards, are in fact shown to be abundantly contaminated. The sheletered microbes, irreversibly anchored on their media, go undetected for the most part by routine investigations performed based on conventional techniques. We developed this topic several times, but this evidence appears, for the time being, to pose too many problems to solve, at least on a psychological and practical level, for being openly accepted. It does indeed require the questioning of a certain number of dogma, the research of an adapted vocabulary, the abandon of outdated techniques and routines no longer corresponding to current scientific concepts. The only advantage of the techniques of the past appears to lie in their capacity to reassure the public opinion by leaning on questionable safety standards. However, as far as industry, environment hygiene, hospitals, and leisure activity areas (in particular, swimming pools), numerous practical applications of release using hydrolases should be considered.

This technique could be applied in decontaminating circuits or water or other liquids. It is well established that microorganisms are solidly attached to the walls of ducts and recipients of all kinds, including those in stainless steel. They proliferate and form agglomerates, whose coats can reach a thickness of 250 to 400 μ. Due to this adherence, germs are sheltered from the action of antiseptics injected in the circuits. This irreversible state accounts for the failures recorded during decontamination attempts.

In 1994, B. Suarez, M. Criado, and their colleagues[308] insisted on the seriousness of the current situation in the foodstuffs industry. Among others, there is the problem frequently posed in hemodialysis circuits in the dairy industry as well as in the foodstuffs and pharmaceutical industry. In another field, with the associations of enzymes and antibiotics, therapeutics have effective weapons available for fighting easily accessible, local infections (ENT, parodontitis, urogenital mucosa). Studies by R. Babin, A. Rigaud, and J. Brisou (1949, 1952, 1975), which are already old, have been widely confirmed in clinical practice.

Unfortunately, none of this is new, but release and the studies devoted to adherence make it possible to understand the molecular mechanisms of these beneficial actions better. The germs released become more accessible to the antimicrobial agents and the defense reactions of the organism. We specified, for example, that hyaluronidase behaves more like a release agent than a "diffusion factor". Its essential action makes the microbes more vulnerable since they are free. Moreover, it makes them sensitive by acting on the parietal elements. Its role in the permeabilization of the tissues thus becomes secondary; it would nevertheless be unfair to question it. It is typical and should be respected.

During our studies, we had the possibility of releasing *Mycobacterium lepreae* of human and experimental lepromas. The fragments treated using hydrolases are clearly richer in bacilli than the controls. The seeding on common media enabled survivals of the Hansen bacillus for several months and positive reseedings. We believe that this is a new approach likely to lead to the successful culture of this bacillus hoped for so long, based on the principle that a released germ cultivates more easily than a stuck bacteria. This is precisely the case with *Mycobacterium lepreae*. The goal is to destroy the magma that protects it and to learn more about the structure of it in order to call on the most suitable hydrolases. In the end, this technology lends itself to a myriad of applications; it calls for the continuation of experiments, subject to the variation of the modalities. The results suggest possible extensions in numerous fields explored by microbiologists.

LET'S CHANGE THE VOCABULARY!

It is recommended that specialists develop a new epistemology of microbioecology. Unicellular beings constitute a very special world in which the language inherited from naturalists and botanists is no longer suitable. The terms "microbial flora", "buccal flora", "intestinal flora", and the curious "vaginal flora" could disappear from the specialized literature without doing any harm. It would be normal to accept the more conformist vocabulary suggested by scientific ecologists, as it is more adapted to the evolution of the concepts. Microbial populations, or communities, constitute "microbiocenoses" and "microbisms". One needs to keep up with the times in every subject. The relentless determination in the outdated application of the language of botany to microbiology leads to confusion.

An isolated cell is in danger, a genuine "heretic of existence", to use the expression of a modern philospher discussing solitude. Unicellular beings are bound to community life, to the formation of agglomerates, to the construction of real fortresses and battlements, at the time of the colonization of all the inert or living interfaces. They find their food and shelter there and the possibility of reproduction. It is only normal for them to benefit as much as possible from these interfaces. Microbiology is a biology of "crowds"; it should be thought of in terms of categories different from those of Linné and expressed in a vocabulary in compliance with the basic data of current science.

When all is said and done, we only know about a small part of this microbiology of the environment, of nature, and of living beings. Although the task is difficult, nothing is materially opposed to carrying out the research programs suggested here which we strongly hope to see undertaken.

A WISH

The experiment results reported here confirm the acknowledged inadequacy of the methods currently in use. The health supervision of resort areas and shellfish farms are striking examples. Water analyses alone only provide partial data that never correspond to reality. For a long time we have been suggesting the control of sand and sediments, preferably after the possible release of microorganisms, with the knowledge that these materials constitute huge reservoirs of microorganisms.

We developed methods enabling the measurement of the enzymatic activity of these microbiocenoses, which is shown to be proportional to the biomass (Yen, 1977).[312] Some reactions make it possible to suspect the presence of undesirable bacteria and, in this case, analyses are continued using releases, isolations, and identifications of the possibly pathogenic germs. Conventional cultures, immunofluorescence marking, or even PCR surveys are implemented, as in the past — the qualitative aspect must constantly prevail over the quantitative aspect. We are not excluding them, but we are aware of the inaccuracies.

It is acknowledged that all living beings house a considerable mass of microorganisms and that nature, from life's infancy, is overloaded with them. Microbes contribute, as a whole, to the balance of the biosphere on all levels. The immune systems of living beings, as marvelous as they are complex, constantly make sure they are protected from aggressions. Thanks to this natural or acquired immunity, humans, like many animals, have escaped from epidemics and deadly epizooties.

"Infectious diseases are the fatal companions of our life" wrote Charles Nicolle[309] in 1933, in the book entitled *The Fate of Infectious Diseases*. The discovery of microbial antagonists and a new generation of vaccinations have profoundly modified epidemiology and prolonged life, but we know that microbes also adapt and have means of resistance. Some even specifically destroy immune systems, which is indeed alarming. It is hoped that health supervision of the environment and food and microbioecology, in general, will adopt other methods of supervision and take more account of the knowledge acquired over the last 30 years, as far as microbes in nature are concerned. Their mechanisms and stages have been rapidly put forward in order for a few lessons to be drawn.

The strategy we suggest is not technically complicated. It is within the reach of any normally equipped laboratory. Nevertheless, experience has convinced me that questioning official and international "Health Standards" remains a delicate topic. Those in charge of Public Health, in most countries, express reservations about anything new. Changes in methods, techniques, habits, or equipment are accepted with difficulty. Wisdom therefore calls for patience. We hope that different teams will have the opportunity to experiment the approach we have outlined, in complete objectivity and total independence. This strategy lends itself to all possible variants and stimulates the imagination. The results recorded will determine its future: accepted, modified, criticized, or rejected. The answer is not mine to give.

BIBLIOGRAPHY

BIBLIOGRAPHY

1. Duclaux, E., *Traité de Microbiologie*, Baillière, Paris, 1898.
2. Alexander, M., *Microbial Ecology*, John Wiley & Sons, London, 1971.
3. Gould, G. W. and Corry, E. L., *Microbial Growth and Survival in Extremes Environments*, Academic Press, New York, 1980.
4. Brisou, J., Mise en evidence des formes masquées d'*E. coli, C.R. Soc. Biol.*, 161, p. 901, 1969.
5. Roth, J. L., *The Lectins Molecular: Probes in Cell Biology and Membrane Research*, Fischer, Iena, 1978.
6. Sutherland, I., The bacterial wall and surface, *Process Biochem.*, 4–8, 1975.
7. Costerton, J. W., Irwin, R. T., and Chang, W. T., Comment collent les bacteries, *Pour la Science*, 5, p. 100, 1978.
8. Costerton, J. W., Irwin, R. T., and Chang, W. T., The bacterial glycolalyx in nature and disease, *Annu. Rev. Microbiol.*, 351, p. 299, 1981.
9. Mazliak, P. , *Les Membranes Protoplasmiques*, Doin, Paris, 1971.
10. Danielli, J. F., and Ridcliford, A. C., *Surface Phenomena in Chemistry and Biology*, Pergamon Press, New York, 1958, p. 246.
11. Paysant, M., Bitran, M., and Wald, R., Polonovski, J., *Bull. Soc. Chim. Fr.*, 52, p. 1257, 1970.
12. Kavanau, J. L., *Structure and Function in Biological Membranes*, 1, Holden Day, Oakland, 1965, p. 760.
13. Vosbeck, K. and Mett, H., Bacterial adhesion influence of drogue, in *Medical Microbiology*, Academic Press, New York, 1983, 3.
14. Maddock, J. R., Alley, M. R., and Shapiro, L., Palized cells, polar action, *J. Bacteriol.*, 1, p. 7125, 1993.
15. Deneke, C. F., Thorne, G. M., and Gorbach, S. L., Attachment pili from enterotoxigenic E. coli pathogenic for humans, *J. Infect. Immunol.*, 26, p. 362, 1979.
16. Deneke, C. F., Thorne, G. M., and Gorbach, S. L., Serotypes of attachment pili of enterotoxil E. coli, *J. Infect. Immunol.*, 32, p. 1254, 1981.
17. Lips, A. and Jessup, N. E., Colloidal aspects of bacterial adhesion, in *Adhesion of Microorganisms to Surfaces*, Ellwood, D. C., Melling, J., Rutter, P., Eds., Academic Press, New York, 1979, p. 5.
18. Nestor, J. and Brisou, J., Incidence sanitaire de l'adsorption des enterovirus sur les sediments, *Rev. Epidemiol. Santé Publique*, 34, p. 181, 1986.

19. Denis, F., Dupuis, Th., Denis, N. A., and Brisou, J., *J. Francais d'Hydrologie*, 8, p. 22, 1977.
20. Shuval, H. T., *Developments in Water Quality*, Shuval, Ann Arbor, 1977.
21. Winogradsky, S., *Microbiologie du Sol*, Masson, Paris, 1949, p. 68.
22. Derjaguin, B. V., *Physico-Chimica Acta*, 14, 1941, p. 633.
23. Verney, E. J., and Overbeek, J., *Theory of the Stability of Lyophobic Colloids*, Elsevier, Amsterdam, 1948.
24. Hill, M. J., James, A. M., and Maxted, W.R., *Biochem. Biophys. Acta*, 63, p. 264, 1963.
25. Rogers, H. T., The bacterial cell wall, *J. Gen. Microbiol.*, 32, p. 19, 1963.
26. Garrett, A. J., *J. Biochem.*, 95, p. 6C, 1965.
27. Lengeler, J. W., La nage des bacteries, *La Recherche*, 217, ch. 21, 1990.
28. Watts, J. W., Dawson, J. R., and King, J. M., The mechanism of entry of virus in plant protoplasts, in *Adhesion of Microorganisms Pathogenicity*, Ciba Found. Symp., 80, Pitman Med., 1981, p. 56.
29. Wilkinson, S. G., Composition and structure of bacterial polysaccharides, in *Surface Carbolydrates of Prokaryotic Cells*, Sutherland, I. W., Academic Press, New York, 1977, p. 77.
30. Reters, M. L., Heidelberg, J., Trowbridge, S., and Kornfeld, S., *J. Bacteriol. Biol. Chem.*, 218, p. 990, 1982.
31. Dazzo, F. B., Kijne, J. W, Haahtela, K., and Korhonen, T. K, Fimbriae, lectins and agglutinins, in *Microbial Lectins and Agglutinins*, Mirelman, D., Ed., John Wiley & Sons, London, 1986, ch. 11, p. 237.
32. Bader, C. and Monet, J. D., *Sté. Chimie Biologique-Colloque*, 6, 1978.
33. Stillmark, H., *Uber rizin ein giftiges ferment aus dem samen von Ricinus communis*, Thesis, University of Dorpat, 1888.
34. Guyot, G., *Zentr. Bakt. Abstr. I. Org.*, 47, p. 640, 1908.
35. Kauffmann, F., *Enterobacteriaceae*, Ejmar Munksgaard public, Copenhagen, 1951.
36. Boyd, W. C. and Shapleigh, E., *J. Immunol.*, 73, p. 226, 1954.
37. Boyd, W. C. and Reguera, B. M., *J. Immunol.*, 62, p. 333, 1949.
38. Erlich, P. , *Dtsch. Med. Wochenschr.*, 17, p. 976, 1891.
39. Erlich, P. , *Fortschr. Med.*, 15, p. 41, 1897.
40. Landsteiner, K., *Wien. Klin. Wochenschr.*, 14, p. 713, 1901.
41. Bruylants, M. and Venneman, M., Le jequirity, *Bull. Acad. R. Med. Belg.* XVIII, p. 3, 1884.
42. Dieuaide, C., *Place Actuelle et Perspective d'Avenir des Lectines Dans le Diagnostic*, Thèse, Faculte de Medecine, Limoges, 1985.
43. Cummings, S. and Kornfeld, S., *J. Bacteriol. Chem.*, 257, p. 1235, 1982.
44. Crane, M. J. and Dvorak, A., *Mol. Biochem. Parasitol.*, 5, p. 533, 1982.
45. Kishida, Y., Olsen, B. K., Berg, R. A., and Prokop, D. S., *J. Cell. Biol.*, 64, p. 331, 1975.
46. Lis, H. and Sharon, N., *Annu. Rev. Biochem.*, 42, p. 541, 1973.
47. Nicolson, G. L., *Int. Rev. Cytol.*, 39, p. 89, 1974.
48. Duguid, J. P. , Smith, I. W., Dempiter, G., Edmunds, P. N., *J. Pathol. Bacteriol.*, 70, p. 335, 1955.
49. Duguid, J. P. and Gillies, R. R., *J. Gen. Microbiol,* 15, p. 6, 1956.
50. Duguid, J. P. and Old, D. C., *Bacterial Adherence*, Chapman & Hall, London, 1980, p. 186.

51. Brinton, C. C., Contribution of pili to the specificity of the bacterial surface, in *The Specificity of Cell Surfaces*, Davis, B. D., Ed., Prentice Hall Engle Woodcliff, 1967, p. 37.
52. Ofek, I. and Beachey, E. M., Mannose binding and epithelial cell adherence of E. coli, *J. Infect. Immunol.*, 22, p. 247, 1978.
53. Bar-Shavit, Z., Goldman, R., Ofek, I., Sharon, N., and Mirelman, D., *J. Infect. Immunol.*, 29, p. 417, 1980.
54. Sharon, N. and Ofek, I., Mannose specific bacterial surfaces lectins, in *Microbial Lectins and Agglutinins*, Mirelman, D., Ed., John Wiley & Sons, London, 1986, ch. 3.
55. Brinton, C. C., *Trans. N.Y. Acad. Sciences,* (11)27, p. 1003, 1965.
56. Abraham, S. N. and Beachey, E. H., *Advances in Host Defense Microorganisms*, Gallin, J. J., Fauci, A. S., Eds., Raven Press, New York, 1985, 4.
57. Firon, N., Ofek, I., and Sharon, N., *Carbohydr. Res.*, 120, p. 235, 1983.
58. Firon, N., Ofek, I., and Sharon, N., *J. Infect. and Immunol.*, 43, p. 1088, 1984.
59. Sharon, N., Bacterial lectins, in *The Lectins*, Liener, J. E., Ed., Academic Press, New York, 1986, ch. 9, p. 494.
60. Eshdat, Y., Izhar, M., Sharon, N., and Mirelman, D., Flagellar lectin as mediator for adherence to eukaryotic cell surface, *Israel J. Med. Sci.*, 17, p. 468, 1981.
61. Le Minor, L., Tounier, P. , and Chalon, A. M., *Annu. Rev. Microbiol.*,124 A, p. 467, 1973.
62. Goldstein, I. J. and Hayes, C. E., The lectins:Carbohydrate-binding proteins of plants and animals, in *Advances in Carbohydrates Cheminstry and Biochemistry*, Academic Press, New York, 1980, ch. 35.
63. Lis, L. H., and Sharon, N., Biological properties of lectins, in *The Lectins*, Liener, E., Ed., Academic Press, New York, 1986, ch. 5.
64. Pusztal, A., Grant, G., Spencer, R. T., Duguid, T. J., Brown, D. S., Ewen, S. W., Peumans, W. T., Van Damnu, E. J., and Bardoca, S., *J. Appl. Bacteriol.*, 75, p. 360, 1993.
65. Doyle, R. S., Nadjar-Haiem, F., Keller, R., and Frasch, C. E., *J. Eur. Clin. Microbiol.*, 3, p. 4, 1984.
66. Prokop, O., Schlesinger, D., and Rackwitz, A., *Immunol. Allergieforschn.*, 129, p. 402, 1965.
67. Mäkelä, O., *Ann. Med. Exp. Biol. Fenn,* 35, p. 1, 1957.
68. Gallacher, J. T., Carbohydrate binding properties of lectins: nomenclature and classification, *Biosci. Rep.*, 4, p. 621, 1984.
69. Gallacher, J. T., and Voss, E. W., *Immunol. Chem,* 7, p. 771, 1970.
70. Reitman, M. L., Trowbridge, J. S., and Kornfeld, S., *J. Biol. Chem.*, 215, p. 9900, 1982.
71. Brisou, J., Infections chroniques par microbisme sélectionné et substitué, *Bull. Acad. Nat. Med. Paris*, 116, p. 7, 1952.
72. Hugues, C., Hacker, J., Roberts, A., and Goebel, W., *J. Infect. Immunol.*, 39, 1983.
73. Kocourek, J. and Hofejsi, V., *Nature*, 290, p. 188, 1981.
74. Kocourek, J., *The Lectins*, Liener, E., Ed., Academic Press, New York, 1986, ch. 1, p. 3.
75. Dey, P. M., Pridham, J. B., and Sumar, N., The Lectins, in *Phytochemistry*, 21, ch. 6, p. 2195, 1982.
76. Dey, P. M. and Pridham, J. B., *J. Biochem.*, 113, p. 49, 1982.

77. Wu, A. M., *Mol. Cell. Biochem.,* 61, p. 131, 1984.
78. Muller, W. E., Zahn, I., Muller, B., Kurulec, G., Uhlenbruck, G., and Waith, P., *J. Eur. Cell. Biol.,* 24, p. 28, 1981.
79. Sutherland, I. W., The Bacterial Wall and Surface, *Process Biochem.,* p. 4, 1975.
80. Richelle-Maurer, E., and Moureau, Z., Mise en évidence de la formation du glycolalyx in vitro, *Ann. Inst. Pasteur,* 138, p. 693, 1987.
81. Walsby, A. E., Kinsman, R., and George, K. I., The measurement of gas vesicle volume and buoyard density in planktonic bacteria, *J. Microb. Meth.,* 293, 1992.
82. Ramakrishnan, S., Duham, D., Gras, M. J., Ratcliffe, M. I., Van alphen, Z., and Coulton, J. W., *J. Bacteriol.,* 174, p. 4007, 1992.
83. Brinton, C. C., *Trans. N.Y. Acad. Sciences,* 27, p. 1003, 1965.
84. Hull, C. L., Hull, R. A., Minshow, B. H., and Falkow, S. J., *J. Bacteriol.,* 151, p. 1006, 1982.
85. Purcell, M., Eisenstein, B. I., Ofek, J., and Beachey, E. H., *J. Infect. Immunol.,* 37, p. 792, 1981.
86. Eisenstein, B. I., *Science,* 214, p. 337, 1981.
87. Eisenstein, B. I., Microbiology, Schlessinger, D., Ed., *Am. Soc. Microbiol.,* Washington, D.C., 1982, p. 308.
88. Ofek, J., Nosek, A., and Sharon, N., Mannoze specific adherence of E. coli, *J. Infect. Immunol.,* 34, p. 708, 1981.
89. Haber, M. J., Mackenzie, R., Chic, K. S., and Asscher, A. W., *Lancet,* 1, p. 586, 1982.
90. Schwartz, A. L., Fridovich, S. E., and Lodish, H. F., *J. Biochem. Chem.,* 257, p. 4230, 1982.
91. Magrou, E., and Brisou, J., *Cr. Sté Hopitaux de Paris,* 1945.
92. Maayan, M., Ofek, I., Madella, O., and Aronson, M., *J. Infect. Immunol.,* 49, p. 785, 1985.
93. Normark, S., Bajà, M., Goransson, M., Lindberg, F. P. , Lund, B., Norgren, M., Uhlin, B. E., Genetic and Biogenesis of E. coli adhesins, in *Microbial Lectins and Agglutinins,* Mirelman, D., Ed. John Wiley & Sons, London, 1986, ch. 5, p. 113.
94. Orkov, F., and Sorensen, K. B., *Pathologia Scandinavia Acta,* 83, p. 25, 1975.
95. Lindahl, M., Brossmer, R., and Waldström, J., A sialic acid specific haemagglutinin of enterotoxigenic E. coli, in *Lectins,* Borg-Hansen, T. C, Breborowicz, C. Z., Eds., Walter de Gruyter, Berlin, 1985, IV, p. 425.
96. Steidler, L., Remant, E., and Fiers, W., *J. Bacteriol.,* 175, p. 7639, 1993.
97. Normark, S., Lark, D., Hull, R., Norgren, M., Baga, M., O'Hanley, P. , Schoolnik, G., and Falkow, S., *J. Infect. Immunol.,* 41, p. 942, 1982.
98. Parkkinen, J., Finne, M., Achtman, V., Vaisanen, V., and Korhonen, T. K., *Biochem. Biophys. Res. Com.,* 111, p. 456, 1983.
99. Normark, S., Bajà, M., Goransson, M., Lindberg, F. P. , Lund, B., Norgren, M., and Uhlin, B. E., Genetic and Biogenesis of E. coli adhesins, in *Microbial Lectins and Agglutinins,* Mirelman, D., Ed., John Wiley & Sons, London, 1986, ch. 5, p. 113.
100. Evans, D. G. and Evans, D. J., *J. Infect. Immunol.,* 21, p. 638, 1978.
101. Baga, S., Normark, J., Hardy, P. , O'Hanley, D., Lark, O., Olsson, G., Schoolnik, G., and Falkow, S., *J. Bacteriol.,* 157, p. 130, 1984.

102. Gaastra, W. and De Graaf, F. K., *Microb. Rev.*, 46, p. 129, 1982.
103. Mooi, F. R., De graaf, F. K., and Van Embden, J. D., *Nucleic Acids Res.*, 6, p. 849, 1979.
104. Shipley, P. I., Dougan, G., and Falkow, S., *J. Bacteriol.*, 145, p. 920, 1981.
105. Henry, T. J., and Pratt, D., *Proc. Nat. Acad. Sci. U.S.A.*, 62, p. 800, 1969.
106. Norgren, M., Normarck, C., Lark, D., O'Henley, P., Schoolnick, G., Falkow, S., Svanborg-Eden, M., Bàga, M., and U'Hlin, B. E., *E.M.B.O.*, 1, 3, p. 1159, 1984.
107. Van Die, J., Mergen, I., Hoekstra, W., and Bergmans, H., *Med. Gener. Genet.*, 194, p. 528, 1984.
108. Low, D., David, V., Lark, D., Schoolnik, G., and Falkow, S., *J. Infect. Immunol.*, 43, p. 353, 1984.
109. Certes, A., *C. R. Acad. Sciences Paris*, 98, p. 690, 1884.
110. Certes, A., *C. R. Acad. Sciences Paris*, 99, p. 385, 1884.
111. Oppenheimer, C. H. and Zo Bell, C. E., Bears Foundation, *J. Marine Res.*, 15, 1952.
112. Fischer, E., *Untersuchungen uber Kohlen Hydrate und Fermente*, Springer-Verlag, Berlin, 1909, p. 2844.
113. Erbing, C., Kenne, L., Lindberg, B., Lonngren, G., and Sutherland, J. W., *Carbohydr. Res.*, 50, p. 115, 976.
114. Dutton, G. S., *Advances in Carbohydrates Chemistry and Biochemistry*, 30, p. 10, 1974.
115. Thurow, B., Shoy, Y. M., Frank, N., Niemann, H., and Stirm, S., *Carbohydr. Res.*, 41, p. 241, 1975.
116. Harden, A. and Young, W. J., The alcoholic ferment of yeast juice, *Proc. Roy. Soc.*, B77, p. 405, 1906.
117. Beijerinck, M. W., *Proc. Doninkl. Acad. Wetenschap*, 12, p. 795, 1910.
118. Jarman, T. R., Bacterial alginate synthesis, in *Microbial Polysaccharides and Polysaccharases*, Berkley, R. C., Goodway, G. W., Ellwood, D. C., Eds., Academic Press, New York, 1979, ch. 2.
119. Allen, P. Z. and de Kabat, E. A., *J. Am. Chem. Soc.*, 81, p. 4382, 1959.
120. Guggenheim, B. and Schroeder, H. L., *Helvetica Odontologica Acta*, 11, p. 131, 1967.
121. Johnson, M. C., Bozzola, J. J., and Shechmeister, J. L., *J. Bacteriol.*, 118, p. 304, 1974.
122. Robbins, P. W., Bray, D., Dankert, M., and Wright, A., *Science*, 158, p. 1536, 1967.
123. Robbins, P. W., Krag, S. S., and Lavi, T., *J. Bacteriol. Chem.*, 252, p. 1780, 1977.
124. Troy, F. A., Frerman, F. E., and Heath, E. C., *J. Bacteriol. Chem.*, 246, p. 118, 1971.
125. Sutherland, I. W., Bacterial exopolysaccharides, in *Surface Carbohydrate of the Prokaryotic*, Sutherland, I. W., Ed., Academic Press, New York, 1977, p. 459.
126. Svanborg-Eden, C., Hagberg, L., Hanson, T., Leffler, H., and Olling, S., Adhesion of E. coli in urinary tract infection, in *Adhesion and Microorganisms Pathogenicity*, Ciba Foundation Symposium, 80, Pitman Medical, 1981, p. 161.
127. Sonessen, A. and Jantzen, E. J., *J. Microb. Meth.*, 15, p. 241, 1992.

128. Brisou, J., Barret, J., Mahklouf, B., and Morcellet, B., *C. R. du C.I.E.S.M.*, 83-84.
129. Strange, R. E. and Dark, F. A., *J. Biochem.*, 62, ch. 429, 1956.
130. Strange, R. E. and Dark, F. A., *Nature*, 177, p. 186, 1956.
131. Gallacher, J. T., *Biosc. Rep.*, 4, p. 621, 1984.
132. Lindahl, M., Faris, A., Waldstrom, T., and Hjerten, S., *Biochem. Biophys. Acta*, 677, p. 471, 1983.
133. Murray, P. A., Levine, M. T., Tabak, L. A., and Reddy, M. S., *B.B.R.C.*, 106, p. 390, 1982.
134. Rampal, R. and Pyle, M., *J. Infect. Immunol.*, 41, p. 339, 1983.
135. Buchanan, T. M., Pearce, W. A., Schoolnik, G., and Arko, R. J., *J. Infect. Dis.*, 136, p. 132, 1977.
136. Wadstrom, T., Faris, A., Lindahl, M., Hjerten, S., and Agerup, B., *Scand. J. Infect. Dis.*, 13, p. 129, 1981.
137. Kendall, F. E., Heidelberger, M., and Dawson, M. H., *J. Biol. Chem.*, 118, p. 61, 1937.
138. Baddiley, J., Buchanan, J. C., and Carss, B., *Biochem. Biophys. Acta*, 27, p. 220, 1968.
139. Mandelstam, J., and Strominger, J. L., *Biochem. Biophys. Res. Com.*, 5, p. 466, 1961.
140. Lilly, M. D., *J. Can. Microbiol.*, 28, p. 11, 1962.
141. Archibald, A. R. and Coapes, H. E, *J. Bacteriol.*, 125, p. 1195, 1976.
142. Hay, J. B., Wicken, A. J., and Baddiley, J., *Biochem. Biophys. Acta*, 71, p. 188, 1963.
143. Perkins, H. R., *J. Biochem.*, 86, p. 475, 1963.
144. Salton, M. R., *The Bacterial Cell Wall*, Elsevier, New York, 1964, ch. 1.
145. Ghuysen, J. M., Strominger, I. L., and Tippero, I., *Compr. Biochem.*, 26A, p. 53, 1968.
146. Wicken, A. J. and Knox, K. W., *Science*, 187, p. 1161, 1975.
147. Shockman, G. D. and Wicken, A. J., *Chemistry and Biological Activity of Bacterial Suface Amphiphils*, Academic Press, New York, 1981.
148. Slade, H. and Slamp, W. C, *J. Bacteriol.*, 84, p. 345, 1962.
149. Duckworth, M., *Surface Prokaryot Cell*, Sutherland, J. W., Ed., Academic Press, New York, 1977, p. 77.
150. Beachey, E. H. and Sampson, W. H., *Microbial Adhesion Surface*, Berkley, H., Lynch, J. H., Eds., Ellis Horwood, 1980.
151. Beachey, E. H. and Ofek, I., *J. Exp. Med.*, 143, p. 759, 1976.
152. Button, J. D. and Hemmings, N. L., *Biochemistry*, 15, p. 5, 1976.
153. Hansen, J. A. and Norsk, M., *Laegevidensk*, 3K, ch. 4, p. 1, 1874.
154. Hansen, J. A., *Virschow's Arch.*, 79, p. 32, 1880.
155. Asselineau, J., Buc, H., Jolles, P., and Lederer, E, *Bull. Soc. Chim. Fr.*, 128, 1953.
156. Asselineau, J., *Bull. Soc. Chim. Fr.* 135, 1960.
157. Asselineau, J., *The Bacterial Lipids*, Herman, Paris, 1966.
158. Snell, E. E., Radin, N. S., and Ikawa, M., *J. Biol. Chem.,* 217, p. 803, 1955.
159. Lederer, E., *Pure and Applied Chemistry*, Butterworths, London, 1961, 2, p. 587.
160. Ross, A. H. and Connell, H. M., *Biochem. Biophys. Res. Com.*, 74, p. 1318, 1977.

161. Draper, P., *Mycobacterium Leprae*, Symposium Sté Fr. Microbiol., Institut Pasteur, 1984.
162. Gaylor, H. and Brennam, P. J., *Annu. Rev. Microbiol.*, 41, p. 645, 1987.
163. Minnikin, J., *Annu. Rev. Microbiol.*, 41, 1987.
164. Weber, K. and Osborn, M., Les molécules de la vie, *Bibl. pour la Science*, Paris, 1984, p. 89.
165. Edelman, G. M. and Wang, J. L., *J. Biochem. Chem.*, 253, p. 3016, 1978.
166. Svanborg-Eden, C. and Jodal, U., *J. Infect. Immunol.*, 26, p. 837, 1979.
167. Buscher, H. J., Werkamp, A. M., Van der Mei, H. C., Van Pelt, A. W., De Jong, H. P., and Areus, P., *Appl. Environ. Microbiol. J.*, 48, p. 980, 1983.
168. Freter, R., O'Brien, P. C., and Macsal, M. S., *J. Infect. Immunol.*, 34, p. 234, 1981.
169. Le Brec, E. H., Spinz, H., Schneider, H., and Formal, S. B., *Proceeding of Cholera Symposium*, p. 72, 1965.
170. Dubos, R. I., Schaedler, R. W., Costello, R., and Hoet, P., *J. Exp. Med.*, 122, p. 67, 1965.
171. Sugarnam, D. and Donta, S. T., *J. Gen. Microbiol.*, 115, p. 509, 1979.
172. Allweis, B., Dostal, J., Carey, K. E., Edwards, T. F., and Freter, R., *Nature*, 1977.
173. Freter, R., Mechanisms of association of bacteria with mucosa surface, in *Adhesion and Microorganisms Pathogenicity*, Ciba Foundation Symposium, 80, Pitman Medical, 1981, p. 36.
174. Gwynn, M. N., Webb, L. T., and Robinson, G. N., *J. Infect. Dis.*, 144, p. 263, 1981.
175. Rigaud, A., Brisou, J., and Babin, R., Thérapeutiques locales à base de papaïne, *Presse Med.*, 31, p. 722, 1956.
176. Brisou, J., Rigaud, A., and Babin, R., Activité bactériostatique de la papaïne, *Ann. Inst. Pasteur*, 77, p. 208, 1949.
177. Brisou, J., Babin, Ph., and Babin, R., *C. R. Soc. Biol.*, 169, p. 3, 1975.
178. Brisou, J., Denis, F., Babin, Ph., and Babin, R., Action du lysozyme sur les streptocoques D, *J. Med. Bordeaux*, 9, ch. 24, p. 64, 1976.
179. Jones, G. W., The attachment of bacteria to the surfaces of animals cells, in *Microbial Interactions*, Reissig, L., Ed., Chapman & Hall, London, 1977, p. 139.
180. Brady, P. G., Vannier, A. M., and Banwell, J. G., *Gastroenterology*, 75, p. 236, 1978.
181. Freter, R., O'Brien, P. C., and Macsai, M., *J. Clin. Nutr.*, 32, p. 128, 1979.
182. Forestier, C., Darfeuille-Michaud, A., Wabor, E., Rich, C, Petat, E., Denis, F., and Joly, B., Adhesion d'E. Coli responsable de diarrhées aux cellules HEP. 2, *Soc. Fr. Microbiol.*, p. 45, 1988.
183. Birkhead, D., Rosell, K. G., and Bowdem, G. H., *Arch. Oral Biol.*, 24, p. 63, 1979.
184. Rutter, P. , Ellwood, J., and Melling, P. , The accumulation of organisms on the teeth, in *Adhesion of Microorganisms to Surfaces*, Rutter, P., Ed., Academic Press, New York, 1979, p. 139.
185. Swanson, J., *Microbiology*, Schlessinger, Washington, D.C., 1975, p. 124.
186. Swanson, J., Studies of Gonococcus infections, *J. Exp. Med.*, 137, p. 571, 1973.
187. Beachey, E. H. and Ofek, I., *J. Exp. Med.*, 143, p. 759, 1976.

188. Beachey, E. H, Berkley, H., and Lynch, J. H., *Bacterial Adhesion to Surfaces*, Ellis Harwood, 1980.
189. Vincent, R. and Pretet, H., *Paris Médical*, Juin, 1933.
190. Vincent, R. and Lehman, J., *Rev. Stomatologie*, 52, p. 794, 1951.
191. Antoine, E., *Arch. Appareil Digest. Nutr.*, 29, p. 588, 1939.
192. Paoli, A. and Grazziani, P., *Med. Trop.* Octobre, 1948.
193. Moustardier, G. and Brisou, J., *Ann. Inst. Pasteur*, 80, p. 355, 1951.
194. Gibbons, R. J. and Van Houte, J., *Annu. Rev. Microbiol.*, 29, p. 19, 1975.
195. Sonju, T. and Skjorland, K., Pellicle composition an initial bacterial colonisation, in *Microbial Aspects of Dental Caries*, Stiles, H. M., Lorsche, W. I., O'Brien, T. C., Eds., Information Retrieval, London, 1977, p. 133.
196. Boisson, J., Brisou, J., and Brangier, J., *Semaine Hopitaux de Paris*, 53, ch. 14-15, p. 827, 1977.
197. Smith, T. and Taylor, M. S., *J. Exp. Med.*, 30, p. 299, 1919.
198. Sebald, M. and Veron, M., *Ann. Inst. Pasteur*, 105, p. 897, 1963.
199. Fauchère, J. L., Quels sont les facteurs de colonisation de Helicobacter pylori?, *Gastro-Graphies*, Lab. Allard, January, 1994.
200. Belbouri, A. and Megraud, F., *Soc. Fr. Microbiol.*, January, 1988.
201. Jones, G. W. and Rutter, J. M., *J. Gen. Microbiol.*, 84, p. 135, 1974.
202. Smith, H. W. and Huggins, M. B., *J. Gen. Microbiol*, 11, p. 471, 1978.
203. Evans, D. G. and Evans, D. J., *J. Infect. and Immunol.*, 21, p. 138, 1978.
204. Levine, M. M., *Adhesion and Microorganisms Pathogenicity*, Ciba Foundation Symposium, 80, Pitman Medical, 1981, p. 142.
205. Savage, D. C., *Int. Rev. Cytol*, 82, p. 305, 1983.
206. Savage, D. C., *Bacterial Adherence*, Beachey, E. H., Ed., Chapman & Hall, London, 1980, p. 33.
207. Bernard, M. C., *Etude des phospholipases bactériennes*, Thèse, Faculté de Sciences, Poitiers, 1971.
208. Brisou, J., Antibiotiques — Synthèse des complexes B — Sélection en chaînes., *Presse Med.*, 236, p. 862, 1953.
209. Gouet, Ph., *C. R. Soc. Fr. Microbiol. (SFM)*, Sept., 4, 1981.
210. Svanborg-Eden, C., Hagberg, L., Hanson, L. A., Korhonen, T., Leffler, M., and Olling, S., Adhesion of E. coli to urinary tract, in *Adhesion and Microorganisms Pathogenicity*, Ciba Foundation Symposium, 80, Pitman Medical, 1981, p. 161.
211. Eshdat, Y., Ofek, I., Yashouv-Gan, Sharon, N., and Mirelman, D., *Biochem. Biophys. Res. Com.*, 85, p. 1551, 1978.
212. Leffler, H. and Svanborg-Eden, C., *J. Infect. and Immunol.*, 34, p. 920, 1981.
213. Courcoux, P., Archambaud, M., Ouin, V., and Labigne-Roussel, A., *Coll. Microbiol. Clin. (SFM)*, Institut Pasteur, Paris, 1988, ch. 11, p. 5.
214. Schwartz, A. L., Fridovich, S. E., and Lodish, H. F., *J. Biochem. Chem.*, 257, p. 4230, 1982.
215. Giron, J. A., Sukyue, H. A., and Schoolnik, G. K., *J. Bacteriol.*, p. 7391, 1993.
216. Westerland, E., *J. Microbial Meth.*, 13, p. 135, 1991.
217. Byrne, G. L., *J. Infect. Immunol.*, 14, p. 645, 1976.
218. Levy, N. J., *J. Infect. Immunol.*, 25, p. 946, 1979.
219. Tysset, C., Brisou, J., Moreau, R., and Durand, C., *Bull. Ass. Diplomés Microbiol. Nancy*, 123, p. 3, 1971.
220. Boullard, B. and Moreau, R., *Sol; Microflore et Végétation*, Masson, Paris, 1962, ch. 3.

221. Goodman, R. N., Huang, P. Y., and White, S. A., *Phytopathologie*, 66, p. 754, 1976.
222. Sing, V. O. and Schroth, M. N., *Science*, 197, p. 759, 1977.
223. Wallace, A. and Perombelon, M. C., Role of hæmagglutinins in adhesion of Erwinia to potatose tissue, *Appl. Bacteriol. J.,* 74, p. 603, 1993.
224. Lippincott, J. A. and Lippincott, B. B., Receptors and recognitions, in *Bacterial Adherence*, Beachey, E. R., Chapman & Hall, London, 1980, B7.
225. Fahraeus, G., *J. Gen. Microbiol.*, 10, p. 374, 1957.
226. Dazzo, F. B., Yanke, W. E., and Brill, W. J., *Biochem. Biophys. Acta*, 539, p. 276, 1978.
227. Dazzo, F. B. and Truchet, G. L., Interaction of lectins and their saccharide receptors in the rhizobium, *Membrane Biol.*, 73, p. 1, 1983.
228. Dazzo, F. B., Bacterial adhesion to plant root surfaces, in *Microbial Adhesion and Aggregation*, Marshall, K. C., Ed., Springer-Verlag, Berlin,1984, p. 85.
229. Jones, G. W. and Rutter, J. M., *J. Infect. Immunol.*, 6, p. 918, 1972.
230. Marshall, K. C., The effects of surfaces of microbial activity, in *Water Pollution Microbiology*, Mitchell, R., Ed., John Wiley & Sons, London, 1978, 2, ch. 3, p. 51.
231. Burns, R. G., *Adhesion of Microorganisms to Surfaces*, Ellwod, O. L., Melling, I., Rutter, P., Eds., Academic Press, New York, 1979, p. 109.
232. Brisou, J., Moreau, R., and Fernex, M., Intérêt de la méthode enzymatique dans l'étude H2S des sédiments marins, *C.I.E.S.M. — XXVIIIe. Congrès*, Monaco, 1982.
233. Lochhead, A. G. and Chan, F. E., Qualitative studies of soil microorganisms, *Soil Sci.*, 55, ch. 5, p. 185, 1943.
234. Guillaume, T., Brisou, J., Lemaire, J. C., and Valensi, G., *Etude de la Corrosion du Cuivre et du Nickel en Milieu Marin*, Colloque sur la Corrosion, Istambul, 1967.
235. Brisou, J., Valensi, G., Constant, H., and Guillaume, T., *Participation des Bactéries à la Corrosion du Cuivre et du Nickel*, 2nd Cong. International Corrosion Marine, Springer-Verlag, Berlin, p. 355, 1958.
236. Brisou, J., Croissant, I., Grimaudeau, J., Guillaume, T., and Valensi, G., Le rôle des bactéries dans la corrosion des métaux, in *Corrosion - Traitements - Protection*, Finition, 1973, 21, ch. 4.
237. Flechter, M., *Adhesion of Microorganisms et Sufaces*, Ellwood, D. L., Melling, J., Eds., Academic Press, New York, 1979, p. 87.
238. Meadows, P. S., Attachment of marine and freshwater bacteria to solid surfaces, *Nature*, 207, p. 1108, 1965.
239. Meadows, S. and Anderson, J. G., Microorganisms attached to marine sand grains, *J. Marine Biol. Ass.,* 48, p. 161, 1968.
240. Dartevelle-Moureau, Z., *Ecologie Microbienne de la Zone Littorale Belge*, Thèse, Faculte de Sciences, Louvain la Neuve, 1975.
241. Brisou, J., Action des polysaccharases sur les microbiocénoses sédimentaires, *Colloque S.F.M.,* October, 1979.
242. Brisou, J., Le débusquement enzymatique des bactéries des sédiments marins, *C. R. Acad. Sciences Paris*, 290, ch. D, p. 1421, 1980.
243. ZoBell, C. E., *Marine Microbiology*, Waltham, Massachusetts, 1946.

244. Kriss, A., Shewan, J. M., and Kabata, Z., *Marine Microbiology*, Oliver & Boyd, Edinburgh, 1962.

245. Brisou, J., Tysset, C., Moreau, R., and Fernex, F., Techniques et intérêt de l'enzymologie des sédiments marins, *Soc, Fr. Microbiol.*, Colloque, 1981.

246. Niaussat, P., Brisou, J., Lafaix, J. M., and Ehrhardt, J. P., Microbiologie d'un lagon marin clos — Pathologies, *Soc, Pathol. Exotique*, 63, ch. 2, p. 160,.

247. Kjellberg, S., Humprhey, B. A., and Marshall, R. C., *Appl. Environ. Microbiol. J.*, 46, p. 978, 1983.

248. Frenel, P., *Microbiologie de l'estuaire de la Loire*, Thèse, Faculté de Sciences, Nantes, 1978.

249. Heukelekian, H. and Heller, A., *J. Bacteriol.*, 40, p. 547, 1940.

250. Rigomier, D. and Brisou, J., Contribution à l'inventaire des polulations bactériennes du plancton, *C. R . Soc. Biol.*, 161, ch. 8-9, p. 1800, 1968.

251. Blavier, J. L., *Etude de Bactéries Anaérobies Isolées du Plancton Marin*, Thèse Faculté de Medecine, Poitiers, 1971.

252. Brisou, J. and Denis, F., Plancton et microorganisms, in *Hygiène de l'Environnement Maritime*, Masson, Paris, 1978, p. 21–109.

253. Moreau, R. and Brisou, J., Identification de bactéries isolées en Méditerranée, *C.I.E.S.M.*, December, 1974.

254. Brisou, J., Moreau, R., and Denis, F., Intérêt des gazes flottées et du plancton dans la surveillance des eaux de mer, *C.I.E.S.M.*, December, 1974.

255. Rigomier, D., Premier bilan population bactérienne hétérotrophe du zooplancton, *C. R. Soc. Biol.*, 161, ch. 3, p. 579, 1967.

256. Dienert, F. and Guillerd, A., *Ann. Hyg. Publique Ind. Soc.*, 18, p. 209, 1940.

257. Carlucci, A. F. and Pramer, D., *Appl. Microbiol. Biotechnol.*, 8, p. 254, 1960.

258. Oppenheimer, C. H., and Zo Bell, C. E., *J. Marine Res.*, 15, 1952.

259. Oppenheimer, C. H., *Symposium on Marine Microbiology*, Oppenheimer, C. H., Thomas, C.H.C., Eds., 1963.

260. Brisou, J. and de Rautlin de la Roy, Y., Populations microbiennes de surface en haute mer, *C. R. Soc. Biol.*, 159, p. 1454, 1965.

261. Brisou, J., Mesures à prendre pour assurer la salubrité du littoral Méditerranéen, *W.H.O. Cahiers Santé Publique*, Genève, 1975, 62.

262. Rigomier, D., *C. R. Soc. Biol.*, 161.3, p. 579, 1967.

263. Campello, F., Brisou, J., and de Rautlin de la Roy, Y., *C. R. Soc. Biol.*, 157, ch. 3, p. 618, 1963.

264. Kaneko, T. and Colwell, R. R., *J. Bacteriol.*, 113, p. 24, 1973.

265. Vinogradova, N. S., *Deep Sea Res.*, 5, 1962.

266. Brisou, J. and Curcier, H., *Bull. Soc. Pathol. Exotique*, 55, p. 1205, 1962.

267. Brisou, J., *C. R. Soc, Biol.*, 161, p. 901, 1969.

268. Webb, M., *J. Gen. Microbiol.*, 3, p. 418, 1949.

269. Pearl, H. W., *Adsorption of Microorganisms to Surfaces*, Bitton, G., Marshall, K. L., Eds., John Wiley & Sons, London, 1980, ch. 11, p. 375.

270. Nivet, A., *Etude de 155 Cas de Septicémies en Milieu Hospitalier*, Thèse, Faculté de Medecine, Poitiers, 1971.

271. Denis, F., Brisou, J., Hoppeler, A., Creusot, G., and De Boissière, A., *Rev. Epidemiol. Santé Publique*, 20, ch. 3, p. 299, 1973.

272. Driessen, F. M., De Vries, T., and Kingma, F., Adhesion and growth of themoresistant Strepto. on stainless steel during heat treatment of milk, *J. Food Prot.*, 47, ch. 11, p. 842, 1984.

273. Stone, L. S. and Zottola, E., *J. Food Sci.*, 50, ch. 4, p. 957, 1985.
274. Deneyer, S. P., Jassim, S. A., and Stewart, G. S., *Biofouling*, 5, p. 125, 1991.
275. Mittelman, M. W., King, J. M., Sayler, G., and Whrite, D. C., *J. Microb. Meth.*, 15, p. 53, 1992.
276. Allison, D. G., Biofilm: Associated exopolysaccharides, *Microbiol. Eur.*, 16, p. 19, 1993.
277. Bendurger, B., Humb Rinnaarts, H. M., Attendorf, K., Alexander, M., Zehnder, J. B., *Appl. Environ. Microbiol. J.*, 59, p. 11, 1993.
278. Oh Mura, N., Kita Mara, K., and Saiki, H., *Appl. Environ. Microbiol. J.*, 59, ch. 12, p. 4044, 1993.
279. Devasia, P., Natarajan, K., Sathyanarayana, D. N., and Ramananda Kao, G., Surface chemistry of Th. ferrooxidans, *Appl. Environ. Microbiol. J.*, 59, ch. 13, p. 4051, 1993.
280. Konetzka, W. A., Microbial of metal transformations, in *Microorganisms and Minerals*, Weinberg, E. D., Ed., Marcel Dekker, New York, 1977, ch. 4, p. 328.
281. Murr, L. E., Tormoz, A. E., and Brierley, J. A., *Metallurgical Application of Bacterial Leaching*, Academic Press, New York, 1978.
282. Brouillard-Delattre, A. and Cerf, O., Le biofilm: réalité industrielle, *La Recherche,* May, 1993.
283. Dunne, W. M. and Burd, E. M., *Appl. Bacteriol. J.,* 74, p. 411, 1993.
284. Carpentier, B. and Cerf, O., Biofilms and their consequences, *Appl. Bacteriol J.*, 75, p. 499, 1993.
285. Sasahara, K. C. and Zottola, E. A., Biofilm formation by L. monocytogenes, *J. Food Prot.*, 56, ch. 12, p. 1022, 1993.
286. Beech, I. B., Gaylarde, C. C., Smith, J. J., and Geesey, G. G., *Appl. Microbiol. Biotechnol.*, 35, p. 65, 1991.
287. Szmelcman, S., *Bull. Soc. Fr. Microbiol.*, 121, p. 90, 1994.
288. Moreau, R., *114e Congrès National des Sociétés Savantes*, 1989.
289. Razin, M. J., Cox, D. J., and Scott, R. L., *Soil Biol. Chem.*, 14, p. 477, 1983.
290. Skujins, J. T., Potgieter, M. I., and Alexander, M., Dissolution of fungal cell walls by stretonycetes chitinase and glucanase, *Arch. Biochem. Biophys.*, 358, p. 364, 1965.
291. Hosegawa, S. and Nordin, N. H., *J. Bacteriol. Chem.*, 244, 5460, 1969.
292. deVries, O. H. and Wassal, J. G., *J. Gen. Microbiol.*, 73, p. 13, 1972.
293. International Union of Biochemistry, *Enzyme Nomenclature*, Academic Press, New York, 1979, .
294. Brisou, J., Tysset, C., de Rautlin de la Roy, Y., Cursier, R., and Moreau, R., *Ann. Inst. Pasteur,* 106, p. 469, 1964.
295. Brisou, J., Denis, F., and Babin, Ph., Babin, R., Action du lysozyme sur les streptocoques, *J. Med. Bordeaux*, 9, ch. 24, p. 1959, 1976.
296. Rigaud, A., Brisou, J., and Coste, J., Application clinique de la papaïne, *J. Med. Bordeaux*, 1952.
297. Brisou, J., Babin, Ph., and Babin, R., Potentialisation des antibiotiques par les enzymes lytiques, *C. R. Soc. Biol.*, 161, p. 3, 1976.
298. Brisou, J., Collagenases et Procollagenases, *Ann. Inst. Pasteur,* 84, p. 463, 1953.
299. Makhlouf, B., Brisou, J., and Stevenino, J., *C.I.E.S.M.*, 1983.
300. Brisou, J., Lecarpentier, L., and Makhlouf, B., *C.I.E.S.M.*, 1983.

301. Brisou, J. and Makhlouf, B., Débusquement enzymatique des bactéries sédimentaires en présence d'antibiotiques, *C.R. Acad. Sci. Paris,* 295, ch. 3, p. 679, 1982.
302. Brisou, J., *Analyse Bactériologique des Beurres, Le Lait,* 4 Août, 1935.
303. Brisou, J., *Les Enterobactéries Pathogènes,* Masson, Paris, 1946, ch. 2, p. 98.
304. Richelle-Maurer, E., *Intérêt et Conséquence de l'Utilisation du Débusquement Enzymatique des Bactéries,* Faculty Sciences Kinsangi, 3, p. 109, 1983.
305. Richelle-Maurer, E. and Brisou, J., Conséquences du débusquement enzymatique des microorganismes, *Bull. Acad. Med. Paris,* 4, p. 487, 1985.
306. Ebiou, E., *Mycobactéries en Milieu Alpin,* Thèse, Faculté de Sciences, Paris-Créteil, 1988.
307. Jeuniaux, Ch., *Chitine et Chitinolyse,* Masson, Paris, 1963.
308. Suarez, B., Ferreiras, C. M., and Criado, M. T., Adherence of psychotrophic bacteria to dairy equipments surfaces, *J. Dairy Res.,* 59, p. 381, 1992.
309. Nicolle, Ch., *Le Destin des Maladies Infectieuses,* Felix Alcan, Paris, 1933.
310. Jannasch, H. W., *J. Bacteriol.,* 95, 722, 1968.
311. Goodfellow, M., Maduromycetes, in *Bergey's Manual of Systematic Bacteriology,* William & Wilkins, New York, 1989, 4, ch. 30, p. 2509.
312. Yen, T. F., *Chemistry of Marine Sediments,* Ann Arbor Science, 1977.
313. Clarke, A. E. and Gellis, S. S., *Biol. Bull.,* 65, 402, 1933.
314. Razin, S., Kahane, M., Banai, M., and Bredt, W., *Adhesion and Microorganism Pathogenicity,* Ciba Foundation Symposium, 80, Pitman Medical, 1981, 80, p. 98.
315. Razin, S., The mycoplasma membrane, in *Organization of Prokaryotic Cell Membranes,* Ghosh, B. K., Ed., CRC Press, Boca Raton, FL, 1981, 1, p. 165.
316. Bredt, W., Feldner, J., and Kahane, I., Attachment of mycoplasmas to inert surfaces, in *Adhesion of Microorganism Pathogenicity,* Ciba Foundation Symposium, 80., Pitman Medical, 1981, p. 3.
317. Whigham, E. A. and Kleinman, R, *J. Immunopharmacol.,* 162, ch. 5, p. 77, 1983.
318. Baddiley, I., *Essay in Biochemistry,* 8, p. 35, 1972.
319. Toon, P., Brown, P. I., and Baddiley, I., *Biochem. J.,* 127, p. 399, 1972.
320. Ganfield, M. C. and Pieringer, R. A., *J. Biol. Chem.,* 250, p. 702, 1975.
321. Brisou, J., *Microbiologie du Milieu Marin.,* Flammarion, Paris, 1955.

INDEX

INDEX

X

Y